D0072447

FAMILY AND SUPPORT SYSTEMS ACROSS THE LIFE SPAN

FAMILY AND SUPPORT SYSTEMS ACROSS THE LIFE SPAN

Edited by
Suzanne K. Steinmetz
University of Delaware
Newark, Delaware

PLENUM PRESS • NEW YORK AND LONDON

Library of Congress Cataloging in Publication Data

Family and support systems across the life span / edited by Suzanne K. Steinmetz.
    p. cm.
    "Dedicated to Marvin B. Sussman in recognition of his contributions to the field of family sociology"—Dedication page.
    Includes index.
    ISBN 0-306-42792-3
    1. Family. 2. Family—United States. 3. Social networks. 4. Social networks—United States. 5. Life cycle, Human. I. Steinmetz, Suzanne K. II. Sussman, Marvin B.
HQ518.F326 1988   88-12470
306.8'5—dc 19   CIP

© 1988 Plenum Press, New York
A Division of Plenum Publishing Corporation
233 Spring Street, New York, N.Y. 10013

Printed in the United States of America

This book is dedicated to Marvin B. Sussman in recognition of his contributions to the field of family sociology

As a result of rapid technological change and streamlined educational programs, contemporary academe is increasingly characterized by ever-narrowing specializations. Therefore the breadth of the scholarly record achieved by Marvin B. Sussman, to whom we dedicate this book, is most notable. On the cutting edge of ideas, Professor Sussman has confronted the new, untried and controversial viewpoints, and challenged the tried and hackneyed.

The authors of *Family and Support Systems across the Life Span* are colleagues, collaborators in research, and former (or present) students. The chapters, which share a common thread-- the intricate intertwining of the family and support systems throughout the life cycle--can be grouped into three themes.

This volume opens with two previously published articles by Professor Sussman which provide a foundation and linkage with the chapters that follow. *"The Isolated Nuclear Family: Fact or Fiction,"* published in 1959, alerted family scholars to the pitfalls of unquestioningly accepting superficial dictum such as industrialization and the trend towards nuclear family units produces the dissolution of extended kin ties. The second article, *"Law and Legal Systems: The Family Connection,"* was the address he presented when honored with the Burgess Award. This article illustrates the numerous ways in which laws and the legal system impact on families and the need for family scholars to consider these implications when formulating theory and research problems.

The second theme, theoretical, definitional and research approaches to studying family, is represented by three chapters. Teresa Marciano discusses the limiting perspective that results from exclusively depending on quantitative research to provide information on the human condition. She argues for the excitement and insights that are possible when qualitative aspects are also considered. Barbara Settles reviews six approaches for defining a family and the current and future implications of each of these definitions. Using her experiences of collecting research data while on 'safari,' Linda Matocha describes unobtrusive methods of observing and recording everyday life.

The final theme reflects family-support system interaction across the life-span. James Ramey, in his chapter, describes an instrument which assesses the degree to which a group can be considered a support network. Donald Peters' examination of different modes of delivery of Head Start programs reveals that the long-term effects of early intervention programs might be based more on the ability to which they help parents redefine parenting practices than on their ability to remove children from these environments. Chapters by Hyman Rodman and Irwin Deutscher provide us with information that is contrary to that derived from media images. Rodman challenges the myth that all children who remain at home while the parents work are 'latchkey' children who face a plethora of problems resulting from their parent's lack of supervision. Deutscher argues that the elderly, especially those raised during the depression, may be in a more advantageous position than the YUPPIES in terms of their buying power and support services.

Cultural influences, such as those resulting from social class or ethnic group, and cross-cultural differences are examined in several chapters. Margaret Brooks-Terry uses Sussman's 'option sequence model' to trace the disadvantages faced by first generation college students. She describes the negative impact of the role conflicts and subsequent effect on college and career success, resulting from the different expectations held by parents and the college. Murray Straus uses SIMFAM, to compare power relationships exhibited by family members during a game situation. Using families in Bombay and Minneapolis, he documents the influence of the cultural context of behavior. Elina Haavio-Mannila also examines marital power, specifically the impact of cross-gender relations at work on marital power. Using a sample of Finnish men and women from different occupational groups, she notes that women's marital power as well as their marital satisfaction increase when they are employed in non-gender segregated occupations.

The reciprocal interactions between parent and child over the life span are the basis for four chapters. In a case study of an older returning woman student, Gail Whitchurch takes a somewhat broader approach to examining power and exchange by using an intergenerational transaction approach. She demonstrates that not only do parents' values influence the child, but the child's values have a direct influence on changing parental values. The cycle of abuse theory is utilized in an examination of child and elder abuse from a historic and contemporary perspective in Suzanne Steinmetz's chapter. A combination of quantitative and qualitative data are used to demonstrate that children are likely to treat their elderly parents in a manner that reflects they way they were treated as children. Doris Wilkinson provides further analysis of parent-child relationships in her chapter on the transformation of the mother-daughter bond in the later years. Finally, the households of elders and the type and amount of support provided to elders by kin and non-family is discussed by John Mogey. Consistent with earlier findings, he reports that the support given to elders is provided primarily by family members.

The interface between health care professionals and the family are discussed in two chapters. Marie Haug describes the ageism that is rampant among dentists, nurses and doctors and suggests ways in which the health care professional might better meet the needs of their older clients. Focusing on physicians, Betty Cogswell chronicles the mismatch between the current training received by most physicians--in an institutional setting--and the major part of most physicians' practice--in an out-patient or 'walking patient' setting. Although Haug is primarily concerned with the older patient, Cogswell provides models for making physicians more responsive to the needs of clients and their families that are applicable throughout the life span.

In the final chapter of this volume, we examine the role that social supports, primarily the family, has in reducing mortality. Eugene Litwak and Peter Messeri apply a task specific approach to identify the causes of death that are most likely to be reduced by informal support systems.

The chapters in this volume are not only innovative, many will be considered somewhat controversial. They suggest creative approaches for gathering data, imaginative ways to interpret the findings and visionary strategies for maximizing the efficacy of the family's utilization of support systems. As such, they carry on the tradition established by Professor Sussman's seminal work, "The Isolated Nuclear Family: Fact or Fiction."

Suzanne K. Steinmetz
March, 1988

▓ ACKNOWLEDGEMENTS ▓

There are a great number of individuals whose contributions to this book were invaluable. First I want to thank Eliot Werner, the Senior Editor at Plenum, for his support of this project and John Matzka, Managing Editor at Plenum, for his patience in explaining the intricacies of preparing camera-ready copy and his understanding the need to have a short turn-around time.

Susan Dippold, Michele Reif, and Judy Stier of the Department of Individual and Family Studies provided help with the initial typing of the manuscripts. Natalie Scott Foulsham provided considerable technical assistance in all aspects of the preparation of the graphics. Her contribution is gratefully acknowledged. I also want to thank Bruce Robinson for producing a computer generated chart.

The final drafts of this book were prepared on an IBM-AT computer and Hewlett Packard LaserJet. I am most thankful and appreciative of the substantial assistance provided by the staffs of the Micro-computer Resource Center and Hardware Maintenance. They rendered on-the spot-technical assistance and moral support when I experienced an amazing set of computer problems as the deadlines approached.

However, this book would not have been completed without the diligent editorial assistance and proof-reading provided by Beth Cohen. I owe Beth my deepest gratitude for working beyond the call of duty on what must have appeared to be an unending number of drafts.

Finally, I wish to thank the University of Delaware for their generous support of this project.

▓ CONTENTS ▓

xi

✳ 1 ✳

THE ISOLATED NUCLEAR FAMILY: FACT OR FICTION

Marvin B. Sussman

INTRODUCTION

Current family theory postulates that the family in American Society is a relatively isolated social unit. This view of the family stems largely from theories of social differentiation in more complex societies.

A neolocal nuclear family system, one in which nuclear families live by themselves independent from their families of orientation, is thought to be particularly well adopted to the needs of the American economy for a fluid and mobile labor market. It is also suggested that differences in occupational status of family members can best be accepted if such individuals live some distance from each other (Parsons, 1943; Parson & Bales, 1955). Support for these theories is found in the high residential mobility of Americans: one in five families makes a move during a given year and presumably these families are nuclear ones. The existing patterns of occupational and social mobility underlie this movement. It may be said that these mobility patterns "demand" a type of flexible and independent family unit.

The extended American family system, for the married person, consists of three interlocking nuclear families; the family of procreation, the family of orientation, and the one of affinal relations (in-laws) whose interrelationships are determined by choice and residential proximity and not by culturally binding or legally enforced norms. The isolation of these nuclear related families from one another is given further support in current conceptualization of family socialization patterns. Freudian analysis has stressed the difficulties confronting the individual as he seeks emancipation from the family of orientation and has interpreted many emotional prob-

1

lems in terms of in-law and parent-child conflict.   Child rearing specialists have emphasized that the warmth and affection of the parent-child relationship should not be chilled as a consequence of competitive activity from the grandparents, aunt, or uncle.   There is, it is said, sufficient threat to an already fragile nuclear family structure through sibling rivalry and parent-child differences.   Thus having parents or collateral relatives living in the home or even close by adds additional difficulties to the complicated problem of child rearing.   Parents and their young offspring can therefore presumably attain a high level of functioning as a family if they are unencumbered by the presence of relatives.

Some students of ethnic relations have interpreted the breaking away of first generation members from their immigrant families as a necessary prelude to growth and assimilation into American society. In still another field, many social class theorists have emphasized the fluidity of our class system and the necessity of the individual to be shorn of family and other ties which appear to hinder his upward movement within the class system.   In our values we maintain the son is better than the father, or that he should be, and in the process of becoming somebody, or achieving a higher status, more frequently than not it is necessary to discard former identification (particularly those with parents and kin) for newer and more appropriate ones.

Despite these basic positions there are some empirical indications that many neolocal nuclear families are closely related within a matrix of mutual assistance and activity which results in an interdependent kin related family system rather than the currently desecribed model of the isolated nuclear family.   This development, while not superseding in importance the primacy of the nuclear family, may provide a new perspective with regard to its position. This does not mean that nuclear families of procreation are increasingly living with their kin in the same households: they still live in separate residences but frequently they reside within a community with their kin and engage in activities with them that have significant mutual assistance, recreational, economic, and ceremonial functions.   Nor does this mean that a new kin structure and concomitant activities have emerged.   It is probable that a functioning system of kin related nuclear families has always been in existence.

The suggestion that a re-examination of the position of social isolation of the nuclear family of procreation from members of the extended family is necessary and has come from many sources. Professors Sharp, Axelrod, and Blood in Detroit, Edmundson and Breen in New Orleans, Deutscher in Kansas City, Dotson in New Haven have studied family relations in the urban setting.[1]   From these and the author's 1951 study of intergenerational family relationships (Sussman, 1953; 1954) it became evident that the concept of the atomized and isolated nuclear family is not being substantiated by empirical research.

THE CLEVELAND STUDIES

Data on kin and family relationships in Cleveland were obtained in 1956. Area probability samples were drawn from two census tracts classified by the Bell-Shevky Social Area Analysis method (Bell, 1959) as lower middle class and working class respectively. An adult member of 27 working class and 53 middle class households representing 3.5% of the total households in each area was interviewed.[2] The general analysis of kin and family relationships is based upon the 80 cases. Specific analyses by social class are based upon a comparison of 25 matched working and middle-class family systems.

Information on kin relationships among these Cleveland families was obtained from questions about help and service exchanges, the functions of ceremonial occasions, and inter-family visitation. Help items included caring for children, help during illness, financial aid, housekeeping, advice, valuable gifts, etc. Ceremonial occasions included holidays, birthdays, and anniversaries. Questions of visits between kin included the preparing of get-togethers with relatives who lived in or out of town.

Concerning help and service exchanges, practically all families (100% of the middle class and 92.5% of the working class) were considered to be actively involved in a network of inter-familial help by virtue of giving or receiving one or more items of assistance listed above within a one-month period preceding the interview. This help pattern appears to exist along with high residential propinquity of related kin. Seventy per cent of the working class and 45% of the middle class have relatives living in the neighborhood.[3]

The research findings suggest that modern means of transportation permit relatives to live in scattered areas and still operate within a network of mutual aid and service. In Table 1 are enumerated specific types of help and service found within kin networks.

Help during illness is the major form of assistance provided by members of kin related families. Such assistance was given in 92% of the reported illnesses which occurred among kin related families living in the neighborhood in the twelve-month period preceding the interview. Respondent-parents and respondent-sibling reciprocal patterns do not differ significantly. However, service during illness for members of kin related families living some distance from each other show a different pattern. Quantitative data is lacking on this point but case data indicate that help between distant families is given when a family member is critically or believed to be critically ill or is suffering from a long term illness. In this situation disruption of family routines is expected and a member of the well family, most frequently the middle aged parent, volunteers or is asked to come and help. There are no expectations for help in routine illnesses as found among kin related families living in the same neighborhood.

Table 1.   Direction of Service Network of Respondent's Family
and Related Kin by Major Forms of Help

Major Forms of Help and Service	Direction of Service Network				
	Between Respondent's Family and Related Kin Per Cent[a]	From Respondents to Parents Per Cent[a]	From Respondents to Siblings Per Cent[a]	From Parents to Respondents Per Cent[a]	From Siblings to Respondents Per Cent[a]
Any Form of Help	93.3	56.3	47.6	79.6	44.8
Help During Illness	76.0	47.0	42.0	46.4	39.0
Financial Aid	53.0	14.6	10.3	46.8	6.4
Care of Children	46.8	4.0	29.5	20.5	10.8
Advice-Personal & Business	31.0	2.0	3.0	26.5	4.5
Valuable Gifts	22.0	3.4	2.3	17.6	3.4

[a]Totals do not add up to 100 per cent because many families received more than one form of help or service.

The amount of financial aid, care of children (babysitting), advice, and valuable gifts exchanged between members of kin related families is higher in this sample than found by Sharp and Axelrod (1956) in Detroit.   The trend toward mutual aid is in the same direction.   Differences in magnitude are probably due to differences in sampling and characteristics of the two populations.

Twenty-five middle were matched with 25 working class family systems, using the number of nuclear related families (parent and child) as the matching variable.   That is, if the respondent middle class family was composed of a middle aged couple (parental family) who had two married children, each of whom lived in separate households, then a working class unit of similar composition was selected to complete the matched set.   An effort to match family systems on a second variable, namely, sex of the children in the family of orientation, resulted in too few cases for comparison. Matching on the sex factor as well as age of family members would be very important in the study of specific patterns of help and service between kin related families.   These would include the quantity and type of aid exchanged in connection with movement of the immediate family through the life cycle.   This problem will be investigated in the next phase of a longitudinal study on urban family networks.   For the purpose of establishing the existence of aid networks, the matching technique used in this study is adequate.

The desideratum underlying this matching technique is that similar family systems of two social classes have equal opportunities to develop help and service patterns.   The significance of differences between social classes was determined by the differences between matched pairs and not by differences between the samples.

The test for differences between samples is insensitive to specific differences between paired sets and does not take into account the opportunity variable. If the structure of the family system is important in determining help and service patterns then it must be established as a control in the analysis of these patterns by social class.

In Table 2 are found statistically significant differences in four items of help and service by social class.[4] So, for example, middle class more than working class parental and child families give and receive financial aid: the P value of .004 indicates that differences more extreme than that observed would occur in only four out of one thousand trials if the hypothesis of no difference was true. On the other hand, the P value of .80 indicates that in eight out of ten trials the flow of aid from parent to child would be the same for both social classes. There is no significant difference between classes on the amount of help given or received during an illness of a family member.

Table 2.   Differences in Help and Service Exchanged by
           Social Class

Help and Service Items	Middle Class	Working Class	P Values
Help during illness	Same	Same	.80
Financial aid			
Amount exchanged	More	Less	.004
Flow of aid: Parent to child	Same	Same	.81
Care of children	More	Less	.03
Advice (personal & business)	More	Less	.04
Valuable gifts	More	Less	.007

Controlling for distance between parents and married child's households, middle more than working class grandmothers are called upon to "take care of grandchildren." The latter are more often than the former gainfully employed. Among working class couples there is a tendency to use the available married brother or sister than parents for this service.

Middle more than working class family systems exchange advice and give valuable gifts to one another. The network of giving is between parents and children rather than between young married couples and their married sibling families. These differences probably reflect the economic conditions and educational attainment of members of these two social classes.

Middle class more than working class parental and child families give and receive financial help. The network of giving is between parents and children rather than between young married couples and their married sibling families. The flow of financial aid is from parents to children; with respect to such flow, differences by social class are not statistically significant.

Regular social visits between parent and married child families as well as those between siblings occur mainly between those in the same neighborhood. Differences between social classes are not significant. The median frequency for visits between young married couples and parental families during the 12 month period preceding the interview was two to three times weekly and once a week between sibling families.

Seventy-four per cent of working and 81% of middle class families have large family gatherings at least once a year. Ceremonial occasions such as Christmas, anniversaries, and other holidays are used largely for the gathering of kin who live outside of the neighborhood. Approximately half of the families in both classes have large family gatherings at one of the major holidays such as Christmas, New Year, Easter, Thanksgiving or the Fourth of July. Birthdays and anniversaries, while not as popular for family reunions, are more frequently used by the middle class than the working class family as an occasion for a family gathering.

Data from another Cleveland study on population change in a given geographical area (Sussman & White, 1959) supports the idea that many neolocal nuclear families are closely related within a matrix of mutual assistance and activity which results in a kin related family system.

Recently, large numbers of Southern mountain white and Negroes moved into an area of Cleveland called Hough, containing over 21,000 households. Based on a random sample of 401 households in 1957, it was found that the non-white population has risen from less than 5% in 1950 to 59.3% in 1957. Ninety per cent of the non-white population have lived in their present houses five years or less, while 59.7% of the whites have lived in their present houses for the same period of time. This demonstrates rapid mobility and changeover from white to Negro occupancy. In the midst of this invasion-succession process we find a fascinating network of kin ties. Sixty-seven per cent of the whites and 86% of the non-whites have relatives living in the Cleveland Metropolitan Area. Of these, whites have 34% and non-whites 60% of their kin living in the two square mile Hough area. Eleven per cent of the whites and 19.5% of the non-whites and 19.5% of the non-whites have three or more families of relatives living in Hough.

Moreover, in Hough relatives are sources for financial aid, second only to banks, and for assistance in times of personal trouble second only to clergymen. When first, second, and third sources of assistance are considered, relatives are most frequently sought.

Additional data on inter-family visitation and exchange of services such as child care, indicates an intricate matrix of inter-family activities in Hough. Thus, even in high transitional areas of the central city, social interaction revolves around kin related nuclear families. It is suggested that the evidence of propinquity of kin related nuclear families has been overlooked in researches on the urban family and in such areas as use of leisure time, social participation and population mobility.

DISCUSSION

Data have been presented which indicate that many kin related families live close to one another and are incorporated within a matrix of mutual assistance and activity. It has been suggested in the literature that individuals tend to locate wherever there is the best economic opportunity. This purports to explain, in part at least, why individuals usually establish neolocal residence some distance from their kin. Kin ties are considered least important in the individuals choice regarding a job, the choice being made in terms of a "best for me" ideology. The fact is that the rate of economic mobility may be influenced by kin ties. Moreover, persons seeking a new job may now locate where their kin are already established. In such cases relatives can help new arrivals adjust to the new community and provide many of the services already described. Even new arrivals into a community may soon be followed by their relations. This has certainly been the case of ethnic and racial migrations across the United States, the Puerto Rican movement is a good example. The high residential mobility suggested by Peter Rossi (1955) and others may actually be a large scale movement of families into communities where their kin are already established. The notion of economic opportunity as related to mobility is still well founded but it needs modification in view of the findings that kin ties have far more significance in the life processes of families today than we have been led to believe. Parental support to newly married couples may provide the necessary anchorages for reducing mobility. It can be hypothesized that population mobility is not higher than it is because of parental support and other kin dependencies.

Class differences in the type of help and service exchanged reflect more the differences in life styles than willingness to participate in the mutual aid network. The middle more than the working class are in a better position to give financial help and expensive gifts; and non-working middle-class grandmothers are more likely to be "free" to take care of grandchildren. The offering of advice, most likely to be given by the parent to the child, reflects middle more than working class occupational status and child rearing patterns.

A question can be asked as to why there is today in sociological writings so much emphasis upon the social isolation of the nuclear family. Scott Greer and Ella Kube (1959) ask a similar question concerning the emphasis upon the isolation and anomic state of the urban dweller. The urbanite is said to be dependent upon secondary rather than primary group relationships. This view may exist because of a time lag between urban and family theory and research. It may also reflect a cultural lag between what was believed to be a generation ago (or may actually have been) and what exists today. The writings of such men as Durkheim, Simmel, Toennies, and Mannheim contain early 20th century views of family and social life in a growing urban industrial society. Durkheim's research on suicide indicated weaknesses in family structure and the effects of isolation upon the individual. In no way did he indicate the basic features of family structure which did, does today, and will tomorrow sustain its continuity on through time. In other words, a theoretical view tinted towards the ills of social and family life was implanted and subsequent research sought to ferret out the disorganizing features of social life.

The consequences of this process, using the family as an illustration, have been in abundant researches on "what is wrong with the American family," followed by a series of proposals on what should be done, and a dearth of studies on "what is right with the family." The "non-problem" functioning family, representing the majority of any society, carrying on the many daily tasks necessary for survival has not been the subject of much study. Yet examination of the non-problem family evolves an empirical base upon which there can be established the means for accurate diagnosis, evaluation, and treatment of the problem family.

Implied, therefore, is a revision of the social problem orientation and approach to the study of social systems. The approach suggested is that empirical studies of the "normal" or "functioning" units of a social system accompany or precede those made on the problem units of the system. Once the base of what is functioning is established then it becomes possible to evaluate a social problem and to propose adequate alternative solutions.

The answer to the question "The Isolated Nuclear Family, 1959: Fact or Fiction?" is, mostly fiction. It is suggested that kin ties, particularly intergenerational ones, have far more significance than we have been led to believe in the life processes of the urban family. While these kin ties by no means replicate the 1890 model, the 1959 neolocal nuclear family is not completely atomistic but closely integrated within a network of mutual assistance and activity which can be described as an interdependent kin family system. ▓

Expenses for this study were met by grants-in-aid for research from the Social Science Research Council.

ENDNOTES

1. See, for example Deutscher, 1954; Dotson, 1951; Sharp & Axelrod, 1956. British sociologists (Townsend, 1957; Young & Willmott, 1957) find similar evidence of kin in East London families.

2. The choice of census tracts was based on data reported in Green (1954).

3. A note of caution is called for on propinquity of kin. Respondents were permitted to define their own neighborhood without a spatial limitation. It was deliberately left vague because of the difficulties in arriving at an acceptable definition of neighborhood. Thus, respondents might construe the "neighborhood" to be the city of Cleveland, or a census tract, or even the street in which the house was located.

4. A discussion of the statistical procedure used may be found in Tables of the Binomial Probability Distribution. Washington: U.S. Government Printing Office, National Bureau of Standards, June 6, 1949, pp. v-x.

REFERENCES

Bell, W. Social area analysis. In M.B. Sussman (Ed.), *Community structure and analysis*. New York: Crowell, 1959.

Deutscher, I. Husband wife relations in middle age: An analysis of sequential roles among the urban middle classes. Unpublished report, 1954.

Dotson, F. Patterns of voluntary association among urban working class families. *American Sociological Review*, 1951, 16 (Oct.), 689-693.

Green, H.W. *Census facts and trends by tracts: Cleveland real property inventory*. Special Report, 1954.

Greer, S., & Kube, E. Urbanism and social structure: Los Angeles study. In M.B. Sussman (Ed.), *Community structure and analysis* New York: Crowell, 1959.

Parsons, T. The kinship system of the contemporary United States. *American Anthropologist*, 1943, 45 (January-March), 22-38.

Parsons, T., & Bales, R.F. *Family, socialization and interaction process.* Glencoe, IL: The Free Press, 1955.

Rossi, P.H. *Why families move.* Glencoe, IL: The Free Press, 1955.

Sharp, H., & Axelrod, M. Urban population. In R. Freedman *et al.,* (Eds.), *Principles of sociology.* New York: Holt, 1956.

Sussman, M.B. Family continuity: Selective factors which affect relationships between families at generation levels. *Marriage and Family Living*, 1954, 16(May), 112-120.

Sussman, M.B. The help pattern in the middle class family. *American Sociological Review*, 1953, 18(Feb.), 22-28.

Sussman, M.B., & White, R.C. *Hough area: A study of social life and change.* Cleveland: Western Reserve University Press, 1959.

Townsend, P. *The family life of old people: An inquiry in East London.* London: Routledge and Kegan Paul, 1957.

Young, M., & Willmott, P. *Kinship and family in East London.* Glencoe, IL: The Free Press, 1957.

2

LAW AND LEGAL SYSTEMS: THE FAMILY CONNECTION
1981 Burgess Award Address

Marvin B. Sussman

INTRODUCTION

It has been said often and much better by others that Ernest W. Burgess was a renaissance person. His interests were not highly specialized and focused, as found today among "great" sociologists and psychologists. To the contrary, his interests were varied; and as a person for all seasons, he was claimed by many disciplines.

In the area of urban sociology, he promulgated the concept of the concentric circles of urban development and migration, a theory which he attempted to adapt to the streets and sidewalks of Chicago. The presentation of this notion stimulated a birthquake of theoretical development and research in the field of urban sociology. Those who worked with this concept soon found that it did not apply literally and urbanologists developed interesting modifications. One was that movement of people out of the central city was not concentric but rather dispersal occurring along radial streets. Location was a function more related to socioeconomic status, class position, education, and occupation than to successive waves of immigrants. While he held these research interests in the city and its people, he never really liked living in it and preferred the countryside.

Criminology, the study of crime and criminal justice was another special area in which Ernest Burgess received recognition. He worked with others of the famous Chicago school in doing empirical work and theorizing on the development of gangs, deviant and criminal behavior. Along with Park, Anderson and Shaw, he developed a school of thought that has had profound impact upon later day experts in this specialty area. Yet, Dr. Burgess was hardly a criminal and, in fact, never received even a parking ticket during his lifetime.

In the early days when gerontology was a new and upcoming field with a paucity of investigators or concerned persons, Burgess was already recognized as a star.   In a classic paper, Burgess describes the "roleless" role of the aged person- a phrase to be used repeatedly by succeeding generations of scholars - a condition that describes a sizeable minority of persons over 75 years of age (1952).   Yet, Burgess never lived this role.   He continued a full life with his research and scholarship to the very last breath of life.

The family area claims Ernest Burgess.   His earlier studies on marital prediction and success paved the empirical and theoretical ways for subsequent generations of investigators to follow.   Burgess recognized the importance of the family as a primary socialization group and as a unit of analysis with its own unique interaction system.   Yet, Ernest Burgess himself had never married.

One situation, which is known to relatively few persons but which I will share now, occurred in the early days of my career. Before moving to Case Western Reserve University, I had a Carnegie Foundation grant to spend a year   at the University of Chicago. Mr. Burgess had officially retired from the university when I arrived, but he was working in the Family Study Center, a center which was ably led at that time by Nelson Foote.   Working with Dr. Burgess was Bernard Farber who had recently completed his doctorate.   They were continuing research activities, begun before Dr. Burgess' retirement, revising books, and completing scholarly work, and other activities a person does in the autumn of her or his career.   The year was 1955.

Some seven or eight years later, I had the opportunity and privilege to see Dr. Burgess once again.   I had nominated him for an honorary degree at Case Western Reserve University and this recommendation was accepted.   I was particularly pleased with this action because at Chicago Burgess had been overlooked and perhaps overwhelmed by the exuberance of such colleagues as W. F. Ogburn and Robert E. Park.   These persons had received numerous honorary degrees.   This was to be Ernest Burgess' first one.   He came to Cleveland, Ohio, with his sister with whom he was now living, and both were frail but keen of mind and warm of spirit.   I never saw Ernest Burgess after that.   But in my mind he lives on as a gentle person, a scholar, a first-rate sociologist, and as a very warm human being.

So much for relationships and connections.   The knowledge that Ernest W. Burgess was a generalist rather than a specialist has influenced my own thinking and work since I received my doctorate. My concern about law and the connection of the family with law has been expressed in a number of major research undertakings, but never clearly articulated as a central or focal interest.   In this paper I will indicate how legal concerns and issues are interwoven with a number of my primary researches.   Yet, there is a lack of explicitness in relating law to family behavior in my work.   This may be due to the inadequacies of my professional socialization. Whatever is the reason, the experience of Charlie Brown comes close to how I feel and perceive the connectedness of law, legal, and family systems.

Charlie Brown and Lucy were on a cruise ship and Charlie Brown went to Lucy's stand for psychiatric advice.   Lucy said to Charlie Brown that some people on a cruise ship put their deck chairs on the bow of the ship and look at the sights ahead and their future docking.   Some people put their deck chairs on the stern of the boat and look at sights they have passed.   "Charlie Brown," asked Lucy, "where are you on the cruise ship of life?" Charlie Brown replied, "Lucy, I can't even get my deck chair unfold- ed."   The family field has yet to unfold its deck chair and decide if it will go aft, forward, or stand abeam with law and its endemic legal systems.

SOCIOLOGY OF LAW

There is a well established sociology of law whose philosophical and intellectual roots are found in the writings of St. Thomas Aqui- nas, Kant, Marx, Savigny, Maine, Spencer, Durkheim, Blackstone, Coke, Pound, Weber, Llewellyn, Ehrlich, Gurvithc, Hegel, and Tim- asheff.   The concerns of this sociology is with legitimacy and the legality of enacted rules and the power to command which emanates from these rules (Weber, 1954); law and the structure of society; law as a functionally integrative mechanism which fosters social cohesion and solidarity (Durkheim, 1957, 1964); law as power to harmonize in- dividual and social interests (Pound, 1921, 1924, 1938); law, value and mores (Sumner, 1906); law as a formal means of social control (Ross, 1920; Cooley, 1902, 1909; and Small, 1924); law and its inter- relationships with all forms social relations.

This list of sociology of law issues is fragmentary.   Many occupy the time and energy of behavioral theorists and empiricists. What is fascinating is the focus on the individual or the society. Legal rules to reconcile conflicting free wills are at one polar end of a continuum; and the social character of law, maintaining the interests of the society, is at the other end.   Somewhere along the continuum, other units such as families, households, agencies, organ- izations, institutions, and networks can be placed.   Somehow they do not fit in with historic or contemporary conceptualizations of socio- logical jurisprudence or a sociology of law.

FAMILY AND LAW

The family is involved with law only through the acts of individual members. Law and it companion systems may respond in a manner which affects families, but its response is to the individual. The family is not regarded as a unit to be treated as such. The *individual* is the focus, and often the consequences for the family are ignored. Consequently, family scholars interested in law and the family have focused their work on the effects of law and legal systems behavior on the family. These have been labeled "impact studies." How family behavior influences law and the workings of legal systems is only a beginning sociological endeavor.

It is a tribute to NCFR and the 1981 program chair, Sharon Price-Bonham, for scheduling a plenary session on "The Law Acts on Families: Families Act on Laws." The special *Family Coordinator* Issue of "The Family and the Law" is a compendium of thoughtful pieces (Henley, 1977). The NCFR Family Law focus group and its newsletter is another source of stimulating exchange on the reciprocal relationship of law and families. These and related activities form a critical mass for a new or at least a revitalized family law movement.

Initially, I will discuss the concept of law, presenting a few philosophical bases and historic antecedents. Then I will review studies I undertook which might have been conceptualized differently if I had the knowledge of the origins, meaning, power, and significance of the law in the lives of families and family members. The concluding section treats a few critical issues involving law and family relationships.

CONCEPT OF LAW: NATURAL LAW

There are several conceptualizations and processes of law which guide contemporary legal decisions affecting families. Natural law and statutes enacted by legislative bodies provide the framework for judicial decisions expressed through common and civil laws.

Natural law evolves from the creative process of applying reason to experience (Pound, 1938). It embodies moral ideals which function as essential guidelines for appropriate and conforming behavior and for reconciling in any society the need for stability with the inevitability of ongoing change. Ideals of appropriate social forms, usages, and behaviors function as constraints upon revolutionary change but favor evolutionary transitions, give direction to change, and often lead to change.

Natural law's most basic activity in relation to the change process is in providing lawmakers, judges, and jurists with guidelines for decisions based on legal materials shaped in the historic past.

Hence, there is a continuity and stability in the face of evolutionary change.

The origins of natural law are lost in historic time. Historians and legal scholars, however, attribute to Cicero a definition of natural law and statement of principles which has been used over the centuries (Wilken, 1954). Those who sought and incorporated natural law as the cornerstone of their constitutional government shared one common condition. All were experiencing transitions and transformations in the form and rule of government. In such times, people seek rational, eternal, and universal principles which justify law and order over the arbitrary exercise of power and continuous strife and violence.

Natural law is imbedded in the fundamental teachings of the Stoics who stood for the perfection of nature and a true moral order, faith insociableness, goodness and reasonableness of personkind, and belief in a divine creator and intelligence. Creation was the handiwork of this intelligence designed for noble purposes and according to some plan.

In the publication *De Re Publica* (Sabine, 1950) Cicero provides a definition of natural law:

> *There is in fact a true law-namely, right reason-which is in accordance with nature, applies to all men, and is unchangeable and eternal. By its commands this law summons men to the performance of their duties; by its prohibitions it restrains them from doing wrong. Its commands and prohibitions always influence good men, but are without effect upon the bad. To invalidate this law by human legislation is never morally right, nor is it permissible ever to restrict its operation and to annul it wholly is impossible. . . It will not lay down one rule at Rome and another at Athens, nor will it be one rule today and another tomorrow. . . The man who will not obey it will abandon his better self, and, in denying the true nature of a man, will thereby suffer the severest of penalties, though he has escaped all the other consequences which man call punishment.*

The importance of natural law in the 1980s is that it continues to be part of the judicial process. It is the candle in the darkness of worldwide hatred, conflict, and oppression. It works in the conscience, it elevates, it insists that persons be just, compassionate, merciful, truthful, honorable, and faithful. It is, indeed, the light in the darkness (Wilken, 1954).

Natural law provides guides to behavior; it is a fundamental law based on perceptions of the nature of persons. Therefore, it is imprecise and philosophical and cannot deal with specific situations or complex human relationships. Formal positive law is required. Such law takes the form of statutes enacted by legislative bodies; of contracts, customs, decrees, and judicial decisions. Natural law, however, has both a symbiotic relationship and a complementarity with formal or positive law. When positive laws cannot fit new conditions or life situations, such events can be handled by resorting to natural law. Values, rational and moral considerations are entered. New interpretations and decisions follow.

In our historical period natural law has been used successfully in judicial or common law decisions. Judges use natural law as a body of legal precepts to evaluate the efficacy and justness of positive laws and for the interpretation and application of the rules set forth in the formal law.

Common or judicial law is interpretative of statutory documents and other materials such as constitutions or constitutional texts and customs. Over time such interpretations, performed in the "spirit of the constitution" or some other formal document, became in and of themselves powerful determinants of behavior. These decisions became rules as judges kept in mind the "spirit" of the written constitutional texts and legislative acts. Such formal laws needed content and working rules. This was accomplished by the courts whose judges and functionaries built a system of legal precepts not upon a foundation of authoritative but upon abstract formulas of ideals and the natural rights of personkind, traditional practices and experience.

This brief introduction to law is relevant to the connectedness of law to family structure and behavior. Such connectedness, if not apparent, is covert. The structure, activities, and behaviors of family members are circumscribed, constrained and maintained by legal precepts, systems of natural and formal law, the courts, and their judicial decisions. In addition, agencies with missions mandated by law influence and have impact upon the family's structure and behavior.

INHERITANCE, LAW, AND FAMILIES

These connections will be illustrated in a review of selected researchers covering different family and legal issues. The first is a study of inheritance.

In the mid-1960s I had the opportunity with colleagues to undertake a field study on family inheritance patterns (Sussman, Cates, Smith, 1970). The genesis of this study came from earlier empirical work and theorizing. By 1965 there had been a plethora of studies on the topic of the existence of kin networks, especially in urban environments. The description of such networks and detailing patterns of exchange was deemed sufficient so that the next step was to ascertain the meaning and significance of these exchanges and supports.   Therefore, establishing the veracity of the described exchange patterns between family groups, relying primarily on behavior over attitudes and intentions, would be a logical next step and, from my view, an advance.   If one could establish that the distribution of assets--a conveyance of economic worth--occurs primarily along generational or bilateral lines, then the kin network exchange hypothesis would receive strong support.

In the beginning economic transfers were the primary concern; but once the study was conceptualized and underway, we became enmeshed in the legal processes, those associated with probate and the role of the legal profession in will making and estate administration, trusts, and other activities related to inheritance transfers. A few concepts and relevant findings are presented to illustrate the connectedness of family and law.

The orderly transfer of equity, status, and other resources from one generation to the next is both an implicit and explicit act for the maintenance of the group or society.   In the age prior to the development of large-scale bureaucracies with differentiated functions allotted to specialized institutions and organizations, tribes, kin networks, and moieties effected different patterns of inter-family transfers.   The family, however defined, was recognized as the unit which provided stability in the social order and continuity over generational time.   Society-wide transfers made via social security, welfare and insurance systems were a later-day development and made the family less important in these matters.

From this historic experience of effecting stability and accommodating to change over generational time, there emerged laws of testacy and inheritance.   These reflect the states' interest in maintaining public order, respecting the rights of individuals to express their wishes (identified as testamentary freedom) and establishing the parameters of such transfers so as to enable the family topursue its historic responsibility of care for its members.   The result is a complex system of probate to ensure that justice is served to all relations of the deceased, e.g., family members, creditors, and the state.

In the discussion of the origins of law, the emphasis was on natural law as a basic set of philosophical propositions and legal precepts which often guided the specific development of formal law and legal systems.   Inheritance has its roots in natural law, and over the ages there developed a highly formalized system subject to modification by judicial decisions made by jurists who were adapting to conditions that existed during their age period.

The feelings, attitudes, and perceptions expressed and visualized by the 1234 respondents in this inheritance study mirror historic traditions rooted in the past.   The vast majority of participants in the study had internalized the essential precepts of natural law and would have responded as they had whether or not the statutory laws governing the rules of descent and distribution were in existence or

not. This is an analytic hindsight and has to be taken as such. I
will illustrate.

A will is a legal document conveying title of properties, equities and other resources of a person when he or she dies. This is
called testacy and the testator can exercise freedom to assign to
others what he or she wishes them to have after death. Intestacy
is a condition in which there is no will. The devolution of title to
property, equity and other resources of the individual who dies
without a will occurs according to an existing intestate succession
statute. There is no uniform succession statute, and states vary in
their laws. Presumably, all statutes function as surrogate wills and
should approximate the distribution of assets made by a person who
normally would write a will.

Testamentary freedom is one basic principle of will making and
involves the transfer of assets from one generation to the next. The
freedom to convey title to persons of one's choice is called testamentary freedom.   In exercising this right, will makers did not
depart radically from rules of intestacy. Over 98% of the distribution in our study were to the spouse, with provisions for distribution to children upon the death of the testator's partner.

Probing deeper into the quality, meaning and content of these
distributions, it was found that another concept prevailed, *distributive justice*. Testators acted judiciously and, I might add, merci-
fully, in two ways. First, they attempted to be equitable in their
distributions to heirs and legatees, primarily members of their family
of procreation. Testators usually made an informal accounting, taking into consideration intervivos transfers--the giving of assets
during the lifetime of the testator--something which is profoundly
understudied. I will discuss such "during life" transfers subsequently. The second act was to provide a larger share of the assets to
a particular individual, usually a child who took on major responsibility for the long-term and terminal care of the testator. This
occurred almost exclusively when there was only one surviving
spouse. The disowning of kith and kin was negligible. In the 1980s,
the incidence of such disenfranchisement may be increasing, since
older persons are working out new types of living relationships, such
as cohabitation and group living.

Testamentary freedom is likened to free will, self-control, and
individualism. It exists as an important philosophical notion, but its
practice is very limited. Instead, individual heads of families have
internalized the ancient values of stability and orderly change and
relate more to the concept of serial service and reciprocity than to
testamentary freedom. They assume that they will transfer just like
they they received and, more importantly, that family and kin even
distantly located will assume responsibilities for the orderly shift
from life to death and the necessary tidying of details.

Further evidence of the pervasiveness of natural law and its
emphasis upon justice, mercy, support, concern, and empathy is the

behavior of heirs and legatees under conditions of intestacy. The pattern of inheritance distribution in our society is that upon the death of the testator, where there is a will, the assets are usually conveyed to the surviving spouse. Subsequently these assets are transferred to surviving children and other heirs and legatees at the death of the single spouse. This is called the "spouse-all" pattern. The transfer is initially bilateral and then vertical. It is unlike a system of primogeniture, where transfers occur vertically usually from the head of the household to a designated person, traditionally from father to son.

To a large degree, legal statutes and judicial behavior support the spouse-keep-all pattern of transfer where there is no will. In most states the law reads that the surviving spouse will receive one-third or one-half of the assets, with the remainder to be divided among surviving children or their heirs and legatees. Many states provide a lifetime use of the property for the surviving spouse (Glucksman, 1977).

Where there was no will, children could have taken their fair share of the inheritance according to law. In all the cases of intestacy in our study, children elected not to take their fair share under the statutes of succession but allocated irrevocable rights to their portion to the surviving parent, usually the mother. Two factors were operating: one was the sense of distributive justice, namely, the belief held by children that the surviving parent deserved what was left of the resources after having lost her spouse; the second was a concern for the public interest. The public interest is that individuals and other units such as families not become dependent upon society. The children clearly understand that if they do take a fair share of the estate according to law, they may be forcing their parent into a state of dependency and thus assuming psychological, if not economic, responsibility for the care of that person in the long run.

The courts and functionaries associated with the legal system were urging and influencing such heirs and legatees to consider the public interest as their own. They also pointed to the children's eventual inheritance of the remaining equities and stressed some of the ideals of the common law, namely, empathy, identification, intimacy, and love of one's fellow being, especially a family member.

As it occurs in other legal investigations, researchers are guided and constrained by the workings of the legal system. The omission of intervivos transfers as a key process illustrates this point. Intervivos transfers are those gifts of money and goods which occur between family members during the lifetime of the giver. These are largely gifts from parents to children. Such conveyances are probably more significant in judging the quality and intensity of family relationships than any other reported behavior.

We discovered this component of the exchange and sharing relationship by accident. We were interviewing all heirs and legatees, or those who would stand to inherit under intestate statutes. In the course of such questioning, we discovered that intervivos transfers were going on for many years during the lifetime of the deceased. We were not prepared in the research to make an accurate count or to estimate the importance and amount of such transfers. In some family networks, these transfers were far more important in substance and symbolic meaning than those occurring upon the death of the testator.

This glaring omission was a consequence of the legal system's influence upon our thinking and perceptions. We developed this study within the context of the rules and regulations set forth by a legally ordered inheritance system. The precepts of the system and the endemic structure and processes provided the boundaries within which we thought it appropriate to operate. In this sense we were controlled by the system. If we had perceived a reciprocal rather than a unidirectional model of legal and family systems and had taken a life course perspective, a more powerful test of the family exchange hypothesis would have occurred. We were so awed and imbued with respect for the law and its exactness and rightness that we ignored a simple fact. Members of families more significantly and coherently express their feelings and behaviors related to intimate relationships while alive than at the moment of probate. It is easy to be taken over by a formalism of the system, and being "done-in" in this manner may reduce the significance of the issue being probed and provide a less adequate test of it.

To obtain the value and conditions related to economic transfers and the symbolic meaning attached to goods, heirlooms and other remembrances, a study of intervivos conveyances is an important step. This, coupled with a study of probate, would have produced in-depth knowledge of feelings, processes, and the interactional patterns of members of immediate and kin-linked families. Economic transfers of uncertain amounts occurred during lifetimes of the decedents in our study. At death, respondents of the deceased placed more value on the assignment of goods and heirlooms than on the money and property they received. Heirs and legatees viewed such conveyances as expressing the meaning of the lifetime relationship with the deceased. Economic transfers occurred when the average age of recipients was 55 years and did not radically alter their lifestyles. Their psyches were more affected than their bank accounts. Inheritors viewed money and property transfers as a function of serial reciprocity, an almost automatic transfer of resources from one generation to the next.

Statutory law predominates the probate process. Briefly examined next is common or case law derived from judicial decisions. Marriage contracts are used to illustrate the judicial decision process where court rulings create legal tenets and practices in the absence of statutory law.

MARRIAGE CONTRACTS AND COMMON LAW

Our studies of personal marriage contracts demonstrate the evolution of a body of law which has its origins in the values, traditions, and practices of long-gone generations and is modified to reflect changing political, economic, social, and ideological conditions. A statute enacted in a past time, dealing with the conditions and rules of marriage, is constantly reinterpreted. Pro and con judicial decisions on any conflicting issue are used, and courts of varying status and power review and judge this complex aggregate of reasoned postures. There is an ebb and flow of pro and con decision influenced by current value, political, and normative postures.[1]

Modern day personal marriage contracts are a variant of antenuptial agreements which existed from the beginnings of recorded history (Sussman, 1975). Such contracts stipulated economic settlements, social and familial obligations of the marital partners and other family members.

Premarriage agreements were part of the bag and baggage colonists brought to America from England. In time, the provisions concerned with marital and family relationships were discarded from such contracts and the agreement dealt exclusively with economic arrangements and was largely used in situations where one spouse had been married previously. Courts today view what goes on in the home as a family affair and only become concerned and act if there is abuse. Hence, they refuse to adjudicate marital disputes. The court's interest is the economic provisions.

An antenuptial agreement is a contract entered into by two persons who are marrying in which both of them waive spousal property rights at the death of either partner. In our land the laws pertaining to marriage and family stipulate the conditions under which property, equity and assets will be distributed. The antenuptial agreement enables the partners to "guarantee" that what they own will be distributed to their kin by blood or marriage, largely those individuals who are related from a prior marriage or through family of procreation, such as parents. Antenuptial agreements are primarily intended to protect the inheritance interests of the heirs and legatees of the previous marriage. These antenuptial agreements are called "death agreements" in the legal trade.

These agreements, to waive alimony in the case of divorce and property rights in the event of the dissolution of the marriage, have been viewed by the majority of courts in the various jurisdictions of our land as conducive to marital tranquility and thus as in harmony with public policy. Almost universally, contracts of this kind are held by the courts to be valid. About 5% of the antenuptial agreements have problems in establishing validity, largely over the wording of the contract. In general, one can conclude that waivers of "rights of a widow" of all claims against the husband's estate-- those rights and claims automatically conferred by the fact of the marriage--are agreements that are enforceable.

In Florida in 1962, in the case of Posner vs. Posner Court of Appeals, Judge Swan dissented from the trial judge's opinion that an antenuptial agreement should be voided and the judge be free to set alimony. Swan's opinion, later upheld by the Supreme Court, was that after full and complete disclosure of assets prior to marriage"...such an antenuptial agreement concerning alimony provisions should be upheld as a valid contract" (Fleischmann, 1974). The trend in case law regarding personal marriage contracts, including antenuptial agreements, is to treat them as business contracts and as valid for all heterosexual and homosexual relationships and first or repeated marriages. Contractees need not share a common domicile, and palimony suits are becoming more common. This has been the trend up to 1980.

The courts have interpreted public interest and policy broadly. If the contract is entered openly, characterized by fairness, commitment, responsibility, and just intentions, the validity is usually established. The public interest is not to increase the number of individuals who will be supported by the State.

To date, the courts have sidestepped moral issues. While adjudicating conflicts in cases involving cohabitation of consenting adults, the courts have tended to avoid judgments on moral issues related to the living arrangements and behaviors of adults. The celebrated Lee Marvin case illustrates how the court viewed the conflict using the canons of contract law--was there an agreement, was it by both parties, was it honored, where did violations occur, what were the contributions of the parties to the relationship, and who was the injured party.

The court disregarded the moral question of whether Lee Marvin and Michele Triola were "living in sin" or illegally. It judged the case primarily on the precepts of contract law. The Marvin decision and subsequent ones in other jurisdictions, using the Triola vs. Marvin judgment as "evidence," have made cohabitation essentially no different from marriage regarding legal responsibilities. The consequences are important for families and family watchers. First, there is recognition that there is more than one family form, ones which include relationships not restricted to those of blood or created by the act of marriage. Second, the principle guiding most jurists in recognizing the economic rights vested in the cohabitation relationship is whether the bond formed by the two individuals, although not legal, was intended to be a permanent one and has lasted as long as most legal marriages. Third, in recognizing the rights and responsibilities of cohabitees, the courts are reflecting changing community standards.

Given the anticipated social, economic, and political climate of the 1980s, it is problematic whether the common law will develop a model code to protect the civil rights of the nonmarried. Returning to old statutes and legislating new ones intended to protect the nuclear family of procreation and punish those living in other family forms is more than a possibility. The courts will be pressed to consider the will of the people expressed by well organized pressure groups.

The point is that case law is a powerful force for social and economic change. It fluctuates with the times, attempting to respond and be coherent with reinterpreted or new values and mores of the people. It can use or ignore statutes as issues are adjudicated. It uses more sociologic data than legalistic arguments in the presentation of briefs to substantiate its decisions. It is closest to a reciprocal influence and exchange model having impact on the family by its decisions and receiving realities from families by the expression of their views and behaviors.

LAW AND FAMILY: ISSUES FOR PRACTITIONERS, RESEARCHERS, AND THEORISTS

Visitation

Historically, grandparents have always been involved with grandchildren. This has been more of a cultural practice than a legal right. The statutes and the courts have always given priority to the nuclear family of procreation. Custody or visitation rights involving grandparents or other relatives have always been considered by the courts in relation to the doctrine of "the best interests of the child" (Allen, 1976; Goldstein, Freud, and Solnit, 1973). The common and usual practice is that upon the death of a parent the surviving parent automatically is entrusted with the custody of the child. During divorce proceedings, the court takes into account the best interest of the child but does so on the assumption that if a parent is of sane mind and body and can provide for the physical and psychological needs of the child, then it is in the best interests of the child to be with that parent. It is all too frequently a routine action.

The increasing incidence of divorce has potential negative impact upon the psychological well-being of children. This, happening along with lessened concern for continuing parental responsibility upon divorce and with seeking more self-fulfillment, has substantially increased the importance of the role of grandparents in the lives of children and youth. With the continuing high incidence of divorce, there is every reason to believe that children will remain in need of grandparents.

Traditionally, the courts have been very reluctant to interfere with the biologically formed family of procreation. Parents are all powerful in the eyes of the law. They have complete control over their children and by deliberate planning, whim, or fancy, can deny other people access to their children, including grandparents, uncles, aunts, cousins, and other bilateral and intergenerational kin.

Once a "problem" is brought to the attention of the court, it invokes the interest of the State. The State is a third party in what is otherwise a private transaction. In cases of abuse, divorce, or the commission of torts, the State envokes a doctrine of parens patriae or guardianship to insure the welfare of the child. In so pursuing the best interests of the child, the State can make any decision it wishes. Even though parents have a prima facie right to visit their children, this right can be denied if it is in the best interests of the child.

The large number of cases coming to the courts primarily as a consequence of divorce or abuse has led to the delineation of conditions under which the court will allow persons other than parents to have custody or visitation rights. My concern is primarily with visitation rights. Functioning in terms of the best interests of the child, the state determines visitation rights. Increasingly, courts are recognizing that the child has something to say about visitation rights and will consider the child's wishes in making decisions.

In determining the visitation privileges of grandparents, the courts will take into account the interest of the person who has custody of the child. Most jurisdictions abide by the rule that the right of visitation is derived from the right to custody. Courts tend to favor the expressed wishes of the individual who has custody of the minor child. The judicial doctrine used with this posture is the "compelling interest of the parent." The privacy and authority of the individual and household is to be protected from outside interference. Whether this compelling interest is in consonance with the best interest of the child is problematic. The issue is whether denying grandparents visitation rights is positively or negatively affecting the child's psychological or physical needs. To date, the courts generally have held that grandparents have a moral but not a legal interest in relation to visitation. Courts continue to exhibit their extreme bias against age. They seldom consider grandparents as custodians of grandchildren except in extreme situations.

Keeping in mind that the "best interest of the child" doctrine dominates the thinking of the courts in relation to visitation rights of grandparents, there are a number of determinants of visitation. These include: (a) friction in the home; (b) health of the child; (c) prior residence with the grandparent; (d) biological relationship between the child and grandparent; (e) duration and frequency of existing visitations; (f) the death of one or both parents, divorce of the parents, or stepparent adoption of the child (Allen, 1976).

The compelling interest of the parent will prevail over that of the grandparent if a visit to grandparents produces friction in the minor child's home or creates a rivalry between the child's parents and grandparents for affection and attention. The psychological and physical well-being of the child and custodial parent is considered in denying or encouraging grandparent visitation. The court will readily terminate such visitations if it is demonstrated that the well-being of the individuals is being impaired.

The child's experience of living with grandparents is taken into account by the court in establishing visitation rights. If such so-residence produced warm, affectionate, and meaningful relationships and does not produce animosities or rivalries with the child's parents, then the best interests of the child are served. The courts will favor the continuity of these relationships.

In considering visitation rights, jurisdictions take into account whether the relationship between the grandchild and the grandparent is biological. The tendency is to favor visitation with grandparents who are biologically rather than adoptively related.

The length of time of visitations varies from court to court. The "best interest of the child" doctrine is taken into account. There are cases where the visitation period is very close to being a custody arrangement. It is likely that the court would provide very generous rights of visitation to a grandparent if there is a loving relationship with a grandchild; and if the grandparent served as parent, authority figure for the child and had given of herself or himself physically, mentally, and economically for the benefit of the child (Evans vs. Lane, 1911, reported in Allen, 1976).

Some parents may agree by contract or stipulation to the visitation rights of grandparents. Courts have had to deal with this issue when the parent appeals to the court to revoke the provisions of this contract or stipulation. In these situations the "best interest of the child" doctrine prevails and the courts do not feel committed to these contract provisions. They attempt to work out a harmonious solution and weigh the evidence as to the pros and cons of voiding the contract, weighing whether this would be in the best interest of the physical and mental health of the child.

Divorce is looked upon as a severe disruption of the nuclear family by the courts. Consequently, there is a careful examination of all potential relationships for the child which would promote his or her healthy development. The courts continue to adhere to the traditional doctrine of the compelling interests of parents and have recognized that adoptive parents have the same compelling interest, with the authority and power to cut off all relationships of biological relatives. In effect, stepparent adoptions sever grandparent rights in those instances where the natural parent is divested of her or his rights. The courts hold that there is no reason for a non-parent who held rights to visitation to continue to do so.

In recent cases the courts have re-examined this procedure and some reversals from this traditional doctrine have occurred (Mimkon vs. Ford, 1975, reported in Allen, 1976). The "best interest of the child" notion was invoked. The court questioned the validity of the assumption that adoptive parents, in taking on the rights of natural parents could or should effect an irrevocable cutoff of all relationships between the child and his natural relatives. The court held that if natural grandparents could contribute to the well-being of the child and would not attempt to undermine the authority and discipline of the adoptive parents, then visitation rights should be granted.

For the family sociologist and researcher, the essential issues are what are the best interests of the child. As friends and benefactors of children, can we provide the empirical data for court jurisdictions to use in determining what are the best interests of the child in relation to visitation rights of grandparents and other biological kin and nonrelatives? Can evidence be assembled that would take more forcefully into account what children's wishes and preferences are? Given the earlier maturation and development of children, more and more can speak for themselves at younger ages; yet the majority of jurisdictions hold to the traditional doctrine that the child is a chattel in which parents have property rights. This doctrine is the basis for transferring control and authority to adoptive parents upon remarriage. Is such a doctrine appropriate for the 1980s? The essential question is what evidence have we accumulated that will support or negate the chattel doctrine and the varied case law decisions made by courts during the past 150 years which effect custody and visitation practices today.

Filial Responsibility

Filial responsibility laws exist in 40 states of the union. These are infrequently enforced, with the majority of jurisdictions opting to use public sources for support of indigent persons, particularly the aged. Although not verified by any empirical study, it is believed that absorbing the costs of support of such persons through established caretaking systems is more cost effective. The current trend is to reduce the number of people on public assistance and to encourage self-help vis-a-vis family responsibility to care for its own kith and kin. This suggests that more weight will be given to existing filial responsibility laws, with renewed efforts to obtain favorable court decisions in cases involving children supporting other family members, particularly parents. This will be accompanied with heightened enforcement efforts.

The principal of family responsibility for its own members has roots in the Elizabethan Poor Law enacted in England in 1597. This established a reciprocal support relationship between the parent and legitimate child in the event either become destitute. This responsibility was expanded to include other family members by the statute of 1601. Historians note that the care of the poor had long been in private hands with the principal benefactor being the church. Through its system of monasteries in England it took care of the aged and destitute. Henry VIII expropriated the monasteries, and consequently there was a push to establish the public responsibility for the support of the poor, which was effected by the act of 1601.

The sequence of support for the aged and destitute as it then existed in the seventeenth century became the prototype for our value posture and practice regarding filial responsibility in the twentieth century United States. As in seventeenth century England, the primary responsibility for support was oneself. One was obligated to use existing assets and resources to maintain this independent status. If the person could not provide self-support, it then became the responsibility of the family. If the family undertook this and used up its resources or did not have the necessary funds to provide such support, it then became a public responsibility.

Extensive social insurance programs developed in the twentieth century have vastly reduced the incidence of cases requiring family support. The proposed reduction of federal support for such programs will result in new or modified existing policies regarding filial responsibility. Concerned behavioral scientists and family researchers will have to cope with extremely difficult and challenging conceptual, definitional and measurement problems. It is an area of great controversy and uncertainty, especially in (a) determining the need for support of the dependent person, (b) ascertaining the capabilities of the primary family to provide such support while maintaining its own lifestyle and meeting its own goals as a family, (c) establishing parental dependency, (d) establishing the moral or legal responsibility of filial support, (e) determining what is a fair amount of support from each of multiple obligors, (f) ascertaining a person's ability to pay, and (g) deciding if civil or criminal sanctions should be invoked (Garrett, 1980).

A critical issue is the impact of the goals of filial responsibility for family members. Universally the stated goals are to reduce the public costs for support of the destitute and at the same time strengthen the family by giving it this responsibility. Intra- and interfamily dynamics should be researched under these conditions. For example, if a parent becomes destitute because of the lack of third-party coverage through social security or because of inadequate or diminished personal resources--and this person lives in Maine, which has a strong filial responsibility law and enforcement program--would some parents elect to reduce their standard of living in order not to be an economic burden to their children? If support of the aged person is forced upon the primary family, it is highly questionable whether relationships can be developed which are warm, affectionate, intimate, and characteristic of solidarity. If a primary family begrudgingly accepts this responsibility, the consequences for their well-being--which in most instances is marginal at best--it is highly problematic; and the danger is in the perpetuation of poverty (Levy and Gross, 1979).

To date there has not been public interest in imposing a strong filial responsibility law. The new philosophy of self-help and turning to one's primary group for essential services may remain at the level of philosophy, given the increasing economic burdens families are experiencing in recurrent periods of recession and economic depres depression.   In recent historical times large-scale third-party paying systems like social insurance have been the more favored form of supporting those who are not in gainfully employed positions.   The essential issue is whether the eroding of the bases of support for these programs will make filial responsibility a more attractive area for government intervention, with a rigid enforcement system that parallels the one currently in existence to implement child support.

Clear thinking and research are required to estimate the acceptance, cost effectiveness, and likelihood that a voluntary filial support system would strengthen family ties if backed by a law that induces such voluntary compliance.   The contrast would be a law that effects compliance by criminal, civil, or administrative sanctions.   Does such deterrence maintain the indigent person in independent living along with harmonious family relationships?   The issue of economic burden, family strain, and perpetuation of poverty as a consequence of rigid enforcement of filial responsibility laws needs serious study and evaluation before there is a pell-mell rush to enforce existing laws and make current ones more obligatory. This is an area for creative collaboration between legal scholars and family sociologists.

Child Support Enforcement

The current child support enforcement program raises important research questions, policy and ethical issues for the family researcher.   Initiated in 1975, this program intends, with the cooperation of the parental recipient, to locate the noncustodial parent, establish paternity if support obligations are unclear, enforce a support order if it is appropriate and monitor the performance of support payers (Fleece, 1982).   The enforcement program has been deemed to be a success, bringing in $3.27 for every $1 expended in administrative costs (Schossler, 1979).   For the period 1976 through 1980, approximately 5 1/2 billion dollars in child support was processed, almost 470,000 cases of paternity established, and approximately 1.7 million missing noncustodial parents were located. Approximately 600 million dollars of the support pay for children on AFDC was retained by government agencies as reimbursement for welfare expenditures extended on behalf of children (Office of Child Support Enforcement, 1976, 1977).

The genesis of this program was the increasing incidence of children on AFDC.   Compared to 34 of every thousand children in 1950, in 1975 there were 122 children supported on AFDC.   During this 25-year period, two happenings occurred.   First was a value shift: more and more individuals felt less and less guilt about getting help, namely, going on welfare.   Welfare became more of a right to be sought without shame.   It became part of the then

current philosophy of promoting a redistribution of societal resources. The other event was the increased costs of the AFDC
program, coupled with a growing public response that parents were
responsible for their children but, for a variety of reasons, were
shirking this responsibility.

There are a number of important issues to challenge the philosophical and value postures of family scholars regarding the
integrity of the family. The enforcement program requires cooperation of the recipient. If there is noncooperation, a sanction is
imposed; namely, the family is denied funds. While the money is
given to the adult applicant for support of a child, such a denial
puts the child in jeopardy. The child is not given equal protection
with those receiving support because of an act beyond his or her
control.

Cooperation with the state requires the recipient to provide all
types of information that will assist in locating the putative father,
in case the applicant is unmarried; and location of the responsible
adult parent in order to assess and collect child support payments.
The essential issue is how this individual upon whom such demands
are made can be protected legally and psychologically. In states
that still maintain criminal laws against adultery and fornication, the
unmarried AFDC recipient can incriminate herself by such revelations, and it is likely to have a deleterious impact upon her and the
family.

Family scholars have long been concerned about the privacy of
the family. It is a unit which has an inner room well insulated and
protected from the outside world. Identifying sex partners, sub-
jecting oneself to court investigations involving blood testing and
similar practices violates all the canons of privacy. A single parent
in this instance cannot maintain her right to privacy as head of her
family or use this right in making decisions regarding the welfare of
her children.

The search for the putative father is laden with all sorts of
problems for the single-parent family. This individual may be
maintaining ties with the family and can be driven away if pursued
and forced to make financial contributions. He may be doing this in
any case, but if the government is viewed as intruding upon his
rights and efforts to relate to this family, he can be driven off
easily, depriving the family of its immediate emotional, economic,
and social support, and future prospects the single parent may have
for an effective marriage.

There may be many instances in which the unmarried parent is
unwilling to cooperate in finding the putative father because that
individual is a child abuser or even a criminal. She is interested in
shielding her children from this kind of influence. The state in-
trudes on this relationship by "forcing" the unwed mother to reveal
the father and, in so doing, promotes a relationship that does not
portend great stability and harmony for the particular family (Poulin,

1976).

There are many legal, administrative, definitional, and scientific issues being addressed by different publics. The economics are very favorable to maintain this program. It is argued by many that the value of parental responsibility is being re-enforced by this program. Dependency on the state is thereby reduced. The question is whether these changes in attitudes and re-enforcement of values of self-responsibility are being influenced by this program. If they are, at what cost to the psychological integrity of the single-parent family, which now represents approximately 20% of all households in the United States? These are some of the critical issues that need to be addressed by family scholars.

CONCLUSIONS

Family theorists, researchers and watchers should incorporate into their repertoire of conceptual approaches and methodological tools the accounting of law and legal systems in formulating issues and developing problems for family studies. It is critical that we internalize this law and legal system dimension in our diagnostic and conceptual frameworks. We are missing reality, the true state of being, if we examine a behavioral issue or problem and ignore the possible explanatory power which may be attributed to the law and its endemic legal systems.

The very definition of family is multifaceted because of the law. One definition refers to a legal statute which covers marriage and family formation. Other legal definitions emanate from statutes that govern the mission of particular institutions and agencies in the society. As a consequence the welfare and social security systems have a particular definition of family, while local municipalities and communities define family in relation to the education of children in the school, the classification of one- and two-family housing residential areas, and eligibility for locally based human services. As family students, academicians, and clinicians, we should automatically consider family composition, social class, race, age, ethnicity, and other basic characteristics in formulating our studies.

The answer we receive from family members to questions posed in our surveys, clinical- and research-oriented interviews are shaped by the respondent's knowledge and interpretation of the law and regulations governing the particular situation. A welfare or SSI recipient, an unemployed worker, public-housing resident, or free-medical-clinic patient will shape responses and control expression of feelings to conform to the legal regulations. The family's economic, health, and social well-being is not to be jeopardized. Such reactions occur even with our appeals that we do not represent any authority and that we want to identify with the families in our studies.

Law, as a basic component of analytic thinking processes, will enable explanations of observable phenomena which heretofore seemed unexplainable or, if understood, not acceptable. In the human service area, there is a continuous flow of legislation derived from particular social and political policies. The administration of these laws is based on developed regulations--often at odds with the intent of the legislators--which bind the behavior of those responsible for implementation of the law. Bureaucratic functionaries are limited in their behaviors by such interpretations of the law. It is hard for us who have grown up in a humanistic tradition to understand this seemingly inhuman behavior on the part of those whose duty it is to serve others. A related issue is the development of informal systems to counteract this legalized formal structure. An understanding of some of these mechanisms enables one to "case" an institutional system and ascertain quickly the complementarity between the formal and informal systems.

The role of common law in codifying and supporting life-ways over historic time has been grossly ignored and unused as a body of knowledge for understanding the behavior of family members. It is the most pliable body of law and one which we as family students, researchers, scholars and clinicians can most effectively influence in ways consonant with our views of the family's role in complex societies. Presenting research and best clinical practices relevant to family functioning; disseminating such information widely via the media; presenting testimony before public bodies; giving papers, lectures, and other types of presentations can develop a partnership between judicial decision makers and the community of family academics and practitioners.

The directions of this developed complementarity and mutual influencing network will be tracked by the removal of current legal statutes and the passage of new ones that directly bear on the family. A prolife amendment or legislative act taking another form, for instance, will go far beyond defining what is life and the rights of the human being in utero. It also will effect how we perceive and define families, what is appropriate legal family behavior and form of family. The passing of this or similar statutes would have immediate impact on the current pattern of judicial decision making; and over the passage of time such case decisions will reflect the conditions and situations found in society in these later decades.

There is an obvious complementarity between case or common law and statutory law and statutory law, and the directions either of these forms of law take influence the range of issues and problems a researcher or clinician can select to study or treat. These changes in laws affecting families will influence the research process itself. The opportunity to free from the closet sensitive issues or to muzzle the study and treatment of particular family problems is directly related to the themes and content of such laws. Understanding how the separate but complementary legal systems function will save the family scholar and clinician from making mistakes, enabling one to distinguish the impossible from that which *is* possible and to discover perhaps how the impossible may become possible.

The legal system should become a focused area of study by workers in the family field. Such studies should not be left solely to the psychologists, sociologists, historians, and anthropologists of occupational systems. Divorce has become a multibillion dollar industry, and there are enough creative alternative mechanisms for handling the dissolution of marriage--which can be done less expensively and even with less trauma--through disavowal of the adversary process in effecting such separations. This is only one example of the power of the legal system in family matters. The power and influence extends beyond profit taking; it includes interpretation of the law. Because of inexperience and insensitivity to law and how it functions (until there is a problem like separation and divorce or abuse in the family), family members are almost completely controlled. Suggested is consideration of the treatment of law and the family in any course or family training program.

The family and law and its attendant legal systems share a pervasive connectedness. As family academicians and scholars, we have been negligent in our advocacy role relevant to increasing the family's position in relations to the law and legal system and in making ourselves and our constituents aware of the expressiveness and power of the laws and of the ways in which we can influence it. This can be readily corrected by legal- and family-oriented behavioral scholars developing cross-professional and disciplinary research programs. This requires hard work, increasing use of the right side of the brain, and conceptual development of a "legiofamiliobehavioral science," an interdisciplinary approach to understanding human behavior. The yeasting, greening, and harvesting of this wasteland can only occur if such an ecumenical development takes place. ▓

Copyrighted 1983 by the National Council on Family Relations. This article is reprinted from *Journal of Marriage and the Family*, volume 45, No. 1 (February), pp. 333-340. Permission to reprint this article was graciously granted by the National Council on Family Relations.

Expenses for this study were met by grants-in-aid for research from the Social Science Research Council.

END NOTE

1. An excellent treatise on the historical, legal, and social issues of marriage contracts is found in Lenore Weitzman, *The Marriage Contract*, New York: The Free Press, 1981.

REFERENCES

Allen, M.L. Visitation rights of a grand parent over the objections of a parent: The best interests of the child. *Journal of Family Law*, 1976, 15, 51.

Burgess, E.W. Family living in the later decades, *Annals*, 1952, 279, 106.

Cooley, C.H. *Human nature and the social order*. New York: C. Scribner's Sons, 1902.

Cooley, C.H. *Social organization*. New York: C. Scribner's Sons, 1909.

Durkheim, E. *The divisions of labour in society*. New York: The Free Press, 1964 (1st edition, 1893).

Durkheim, E. *Professional ethics and civic morals*. London: Routledge and Kegan Paul, 1957.

Fleece, S.M. A review of the child support enforcement program. *Journal of Family Law*, 1982, 20, 489.

Fleischmann, K. Marriage by contract: Defining the terms of relationship. *Family Law Quarterly*, 1974, 8, 44.

Garrett, W. Filial responsibility laws. *Journal of Family Law*, 1980, 18, 793.

Glucksman, J.R. Interstate succession in New Jersey. *Columbia Journal of Law and Social Problems*, 1977, 12, 253.

Goldstein, J., Freud, A., & Solnit, A. *Beyond the best interests of the child*. New York: The Free Press, 1973.

Henley, L. (Ed.) Family and law (special issue), *Family Coordinator*, 1977, 26(4).

Levy, R.J., & Gross, N. Constitutional implications of parental support laws. *University of Richmond Law Review*, 1979, 13, 517.

Office of Child Support Enforcement. *Annual Report*, (No. 133) U.S. Department of Health, Education and Welfare. Administration of Children, Youth and Families, 1976.

Office of Child Support Enforcement. *Supplemental report*, (No. 155) U.S. Department of Health, Education and Welfare. Administration of Children, Youth and Families, 1977.

Office of Child Support Enforcement. *Annual report*, (No. 113) U.S. Department of Health, Education and Welfare. Administration of Children, Youth and Families, 1979.

Office of Child Support Enforcement. *Annual report*, (No. 86) U.S. Department of Health, Education and Welfare. Administration of Children, Youth and Families, 1980.

Poulin, N.   Illegitimacy and family privacy: a note on material cooperation in paternity suits.   *Northwestern University Law Review*, 1976, 70, 910.

Pound, R.   *The spirit of the common law.*   Boston: Marshall Jones, 1921.

Pound, R.   *Law and morals.*   Chapel Hill: University of North Carolina Press, 1924.

Pound, R.   *The formative era of American law.*   Gloucester, MA: Peter Smith, 1938.

Ross, E.A.   *Social control: A survey of the foundations of order.*   New York: MacMillan, 1920.

Sabine, G.   *A history of political theory.*   Hinsdale, IL: Dryden Press, 1950.

Schlosser, L.J.   An assessment of IV-D's first four years. *Public Welfare*, 1979, 37,22.

Small, A.   *Origins of sociology.*   Chicago, IL: The University of Chicago Press, 1924.

Sumner, W.G. *Folkways.* Boston: Ginn-Company, 1906.

Sussman, M.B.   Occupational sociology and rehabilitation.   In M.B. Sussman (Ed.), *Sociology and rehabilitation.*   Washington, DC: American Sociological Association, 1965.

Sussman, M.B.   Marriage contracts: social and legal consequences. Plenary address, presented at the 1975 International Workshop on Changing Sex Roles in Family and Society, Dubrovnik, Yugoslavia, 1975. (Copy is available from author)

Sussman, M.B., Cates, J.N., & Smith, D.T.   *The family and inheritance.* New York: Russell Sage Foundation, 1970.

Weber, M.   Max Weber on law.   In M. Rheinstein (Ed.), *Economy and Society.* Cambridge, MA: Harvard University Press, 1954.

Weitzman, L.J.   *The marriage contract.* New York: The Free Press, 1981.

Wilken, R.N.   Cicero and the law of nature.   In A.L. Harding (Ed.), *Origins of the natural law tradition.*   Port Washington, NY: Kennikat Press, 1954.

▓ 3 ▓

THE ART AND SCIENCE OF FAMILY SOCIOLOGY

Teresa D. Marciano

INTRODUCTION

In his Preface to the Third Edition of *Sociology*, Ian Robertson describes the premises of his text. *"The first is that sociology is both a humanistic art and a rigorous science. . . "* (1987, pg. vii). As is typical for introductory texts, a chapter is devoted to describing how and why sociology is a science; its artistic component, on the other hand, is implied rather than elaborated.

The history of sociology makes it clear why the "science" should be so emphasized. In arising out of a desire to understand and solve the social problems of an industrializing age and to gain respectability as an integral field of study, the "art" was deemphasized in favor of the more acceptable, less elusive "scientific method."

It is my contention here, however, that the art of sociology is what brought most sociologists to the field. However, the notion of sociology as an art remains suspect due to a misunderstanding of what the "art" of sociology really is. Our understanding of human society is more linked to the "art" than the "science," and this is reflected in the continuing debate over the essentially false dichotomy between qualitative and quantitative methods. These misunderstandings are based in a forgetting of the dual necessity of inductive as well as deductive determinations, and we are called again and again to the art of our discipline to make it useful as a tool by which human beings understand the great enterprise we call "culture and society."

35

ART OR SCIENCE

The question "what is art?" is far older than the discipline of sociology and its arguments over what it is and ought to be. Those who think of "art" as one-half the totality we call "arts and sciences" hear and reflect a common belief that the two areas are mutually exclusive domains, that their methods and wellsprings differ, that one is essentially "creative" and the other "methodical," that art is personal, singular, original, irreproducible, and the product of random genius, of special insight. Science, on the other hand, while advanced by insight and genius, can and must proceed by the rigorous and methodical accumulation of facts, as unbiasedly as possible, which then can be linked together into systematic understandings to produce theories and generalizations which at their best are predictive as well as descriptive.

Such contrasts between "art" and "science" leave the former to the realm of the painter, sculptor or dramatist, while neglecting the essentials of the word "art" and its place in sociology. It is no accident that "art" is the root of the word "articulate," and that while we commonly use the word to describe speech or the capacity of facile speech, it has another meaning: to join things together so that they fit. Our theories are attempts to articulate the facts we have accumulated and the observations we have made, to form that web of connected understandings which we call theory.

The first definition of "art" is the human ability to make or create things. Human ability, realized in patterns of culture, is the center of sociological concern. Other definitions of art, such as "specific skills and their application," show that "art" is the word that subsumes the work we designate separately as "science."

Where the "art" of sociology becomes suspect, however, is not in the area of applied skills, but in the nature of insight, interpretation, implication, and speculation. We emphasize the hypothesis that is "solid," rarely encouraging young scholars to pursue hypothesis that are creative. Yet creativity and its realization in elegant operational definitions, or in the attempt to study data bases in radically new ways, runs against a variety of discouragements. Perhaps the greatest is that requirements for grant proposals must seem "solid." The fact is that "solid" may not be particularly enlightening, but it does enable one to follow the calcified path to publications and the resultant justification for future grants, with concomitant job and tenure benefits. These qualities were already emerging in the discipline in the 1960s and have, if anything, increased in magnitude (Sussman, 1964). In addition to the financial and career-pattern benefits of "solid," large-scale, quantitative research, it is undeniable that creativity and innovation are essentially acts of great risk, putting student and mentor on the line of judgment by those who hold to "rigorous" science against "ephemeral art."

Yet Lewis Coser, a most respected name in sociology, cautioned that *"Sociology is not advanced enough solely to rely on precisely measured variables."* He went on to say:

> *Training the new generation of sociologists not to bother with problems about which data are hard to come by, and to concentrate on the areas in which data can be easily gathered, will result, in the worst of cases, in the piling up of useless information and, in the best of cases, in a kind of tunnel vision in which some problems are explored exhaustively while others are not even perceived (1975:693).*

Coser was not saying in this address that all quantification is bad and all qualitative work good. He was attacking the substantive vacuum he saw in ethnomethodology as well as that in path analysis. True substance, and the theoretical innovations it makes possible, distinguish creative sociology from "safe," but uncreative, sociological routines. Such a sociology is neither artistic nor truly humanistic.

The most basic and anti-humanistic view of art, particularly of sociology as an art, is that art requires things of us that cannot be taught. Yet what musician or painter, however gifted, has not had to learn the essentials of the art to which a life is devoted? There are mystical qualities to art, as there are to science; the physicists reach every day for new words to convey the wonders they are uncovering.

No physicist, however, is born with the knowledge that one must possess as the baseline to creative physics. In the same way, the art of sociology requires the coursework that has been offered. However to enable the art of sociology to be done, it must also encourage the use of what is learned in innovative ways. What ceases to be "artistic" in sociology becomes formulaic repetition; it moves away from the exciting involvement in human society that drives us to look for patterns, to explain (and marvel) at what humans have wrought.

Much of the art of exploring human relationships has its current existence outside the traditional field of sociology, and particularly of family sociology; it resides instead in the actualities of the various therapies. There is a science behind the therapeutic encounter, but the relational actuality of therapist and client is one that is an art: a special kind of social art that can be learned, trained, and refined.

ART AND SCIENCE

Murray Bowen, who is identified with the origin and consolidation of family systems theory (now known as Bowen theory) de-

scribed the intermingling of art and science in the training of
family therapists. He rejected traditional psychoanalytical techniques
precisely because the successful training of therapists under that
system was so unpredictable; he believed that a systematic understanding of family dynamics, applied by the trained observer-
therapists, could be learned and used to help families cope with
dysfunction (Bowen, 1976). The art of family therapy is the application of skills for bringing to light the negative patterns creating
family unhappiness. The explanations of these now-revealed patterns
suggest ways in which families can break free of them.

Anyone who has worked with Bowen theory knows how revolutionary it becomes in adding dimensions of understanding to a study
of family patterns. It can be applied to explain why some family
members "deviate" while others conform; it can show through the
device of the genogram how certain (dys)functional patterns are
transmitted across the generations of a family's history. The
genogram reveals the process itself of that learning we call "cultural
transmission;" it shows us the subcultural world over time that we
call "nuclear" and "extended" families.

Bowen's insights also broaden the ability of family sociology to
draw upon examples in literature as well as in traditional respondent
sources, to demonstrate processes of power and control in family
systems. In Kafka's *The Metamorphosis*, for example, the Samsa
family is described with great subtlety as well as with great irony.
Remember that the story begins with Gregor Samsa's awakening one
morning to find that he had *"changed in his bed to become a
monstrous vermin."* Chillingly, rapidly, he must conclude, *"It was no
dream."* He is the firstborn, the only son, of a now-incapacitated
father and a fussy mother; he is responsible to provide for them and
for his sister. As the story proceeds to detail the way in which
Gregor, in his beetle form, and his family come to deal with each
other, there are descriptions of how the family lived, what their
days were like before Gregor's metamorphosis. Gregor thinks back
to these days because he is surprised that his father, whom he
thought to be so enfeebled, has become a man of greater energy and
strength since his son turned into a bug. Gregor wonders at the
change:

> *Was this the same man who in the old days used to lie
> wearily buried in bed when Gregor left on a business trip
> . . . and who, on the rare occasions when the whole
> family went out for a walk, on a few Sundays in June and
> on the major holidays, used to shuffle along with great
> effort between Gregor and his mother, who were slow
> walkers themselves, always a little more slowly than they
> . . . and, when he wanted to say something, nearly always
> stood still and assembled his escort around him?* (pp.
> 37-38, Corngold transl.).

Kafka's great art here is the ability to describe a system of power and control exercised by the father, who used a self-pronounced debility to govern the actions of his family in such a way, that any similar family we may see or hear about reveals the same dysfunctional control system by one member over the others. Whether one reads the story as metaphor for the oppressive family, an allegory of love and despair, or as in this Bowenesque reading, the perils and sufferings of the triangled child, we understand many families from the one. Great literature has always taught its lessons inductively, and great sociology has often done the same. The "art" in Bowen theory, as in family sociology, resides in an openness to see what is there; it necessitates a questioning of apriori assumptions, critical self-analysis by the analyst as well as by the family.

Art and Science in Sociology

The need to do this in sociology is not unknown. Marvin Sussman called for this when he challenged the unquestioned acceptance of the "isolated nuclear family's" neglect of kin networks (Sussman, 1970). Sussman emphasized that *plausibility* does not substitute for *actuality*, regardless of how rational and logical a description it provides of the social world. Something can be plausible yet not true of society, he points out, when a system of explanation lacks an empirical base.

And this brings us full circle, since the art of sociology is precisely based on empirical understandings, learned skills for research and explanation, and then that further step, the encounter with the phenomenon under study. That particular kind of empirical encounter, the face-to-face contact with sample subjects, becomes critical to understanding why and how things happen in society when data bases provide no correlational clues.

At the 1987 meetings of the National Council on Family Relations (NCFR) in Atlanta, a number of fine papers were presented which used very large national data bases and in which the presenters sought to answer some very intriguing questions about modern family life. At several of these sessions the presenters, after showing all the correlations they had tried in order to account for certain family characteristics, announced that they were stymied, that they could not, using these bases and methods, account for why things were happening as they were.

Several of the presenters noted that they were tempted to go out and get their own samples and to conduct their own studies to find out what was going on. This happened several times at separate sessions of the NCFR, in statements made by researchers unconnected by personal friendship, let alone collegial status. This was a fascinating occurrence because what they said they wanted to do was to use the traditional sociological arts: to do field work, in order to get the sense of what was going on.

At that same conference there were discussions of requirements for the Ph.D. in family sociology because many doctoral candidates were mature students with family obligations, who wanted to get their degrees in the shortest possible time.   Therefore, they often took only the required, career-safe quantitative courses, and neglected courses on qualitative research that were often optional.

This is not a new problem in sociology; the ascendancy of quantitative over qualitative directions of research in graduate sociology was the subject of a 1966 address by Marshall Clinard (Clinard, 1970).   The consequence that he saw then was the ever-growing distance between sociologists and the world they were supposed to be studying and explaining.   This, in his view, placed the validity of sociological conclusions in doubt; ironically, Clinard also saw this as undermining the scientific respectability of sociology, since as Sussman also noted, direct connection with the empirical base is absent.

If this was true for sociology in 1966, when all its subdisciplines including family sociology were far more likely to enroll "lockstep" students who were younger, had fewer nonacademic responsibilities, and who should have been more likely candidates for sociological risk-taking, how much truer must it be today?   Given the numbers of nontraditional-age students in graduate populations, the situation is, unfortunately, more like to occur today.

Evidence that this is indeed the case comes from a serendipitous source--an interview with Russell Jacoby who in 1987 published *The Last Intellectuals.*   Jacoby, issuing a sharp condemnation of the state of American academe, described his surprise at how quickly his own "rebel" Sixties generation had settled into the professions.   We must remember that the generation to which he refers made one President's reelection untenable (Johnson) and unseated the next (Nixon).   It is this same generation that changed the American perception of patriotism from a willingness to die for one's country, to a demand that all be given a chance for a just life.   Of that generation Jacoby says:

> *I wasn't expecting them to transform the world, but I'd by happy if there were, say, a critical, forceful sociology. Instead, there are 5,000 sociologists out there who only talk to each other* (Bernstein, 1987).

To demonstrate the truth of his words, give any undergraduate or educated lay person a copy of *Street Corner Society* or *Blue Collar Marriage*, and then give them a standard article from the *American Sociological Review* or even the *Journal of Marriage and the Family.* Observe how well the information in each source enhances the readers' ability to make sense out of their own lives. Note also that the closure of a discipline on itself is one problem which only begins to touch the question of what is absent by neglecting the findings of other disciplines, such as communication research, social psychology, and social history.

The Implication of Emphasizing Quantification

Irony compounding irony, sociologists not only are aware that this is happening, but two recent successive issues of *The American Sociological Review* contain articles addressing the less-likely success of qualitative articles for publication in prestigious journals (Bakanic, McPhail & Simon, 1987; Grant, Ward & Rong, 1987). The first of these articles examines the manuscript review and decision-making process for *ASR* for the years 1977-1981. The authors note that among the variables examined, "*the sole manuscript characteristic yielding a statistically significant effect was the use of qualitative data analysis. The relationship was negative*" (Bakanic *et al.*, 1987:638). (It is of course impossible to ignore the quantitative demonstration of qualitative disadvantages. Perhaps the complementarity of the two methods becomes more evident, after the inevitable chuckle.)

The second article is a research note which examines gender and methods in sociological research in ten sociology journals for the years 1974-1983. An excerpt from the abstract is in order here:

> *Most articles have been quantitative, but female authors have used qualitative methods more often than males. Writing about gender increased rather than decreased the likelihood of having used quantitative methods for both women and men. We suggest that papers focusing on gender and also using qualitative methods represented double nonconformity and hence were unlikely candidates for publication in mainstream journals* (Grant *et al.*, 1987: 856).

The issue of gender studies and qualitative methods is one that is familiar to feminist scholars, who have shown the link between the "prestige" of quantitative methods and the androcentric bases of that prestige (e.g., Rosenberg, 1982, pp. 238-246). It is probably worth a study of the parallels between the "revolutionary" discipline of sociology, doing field studies of how different social classes lived, or how people entered and were socialized to different (including marginal) occupations, and the revolution that was the Second Women's Movement and the early emphasis by its scholars on the value and worth of qualitative/descriptive research. The fact that quantitative research is now also a women's domain, may be the most convincing evidence of the persistence of old power patterns in academic settings, despite gains made by women in sociology over the past twenty-five years.

The motto (unspoken but clear) that students quickly learn in their graduate work, then, is:

> *If it has no numbers (and varieties of sophisticated cor-*
> *relations, set out in tables expressing frequencies and*
> *degree of significance), it isn't science and is unlikely to*
> *be good sociology.*

This unfortunate view, to the great extent that it seems to exist, is based in a confusion of what is empirical and what is numerical.   Empirical study can lead us to understand process. Numerical expressions of social phenomena are merely a beginning of the sociological detective work needed to figure out why social behavior patterns fit into the numbers that are obtained.   Neglect of the firsthand study of process, especially of the most compelling sort of qualitative research, ethnography, is encouraged by the biases of graduate training, and also by the sheer amount of time and personal involvement required to study any group (or community) firsthand.   Our enduring classic, William F. Whyte's *Street Corner Society,* a grand model for all the ethnography that followed, has had far too few heirs.   Yet, whether done historically as Carden did for the Oneida Community (1969) or more recently in ethnographic/community studies as Schwartz (1987) has done, one is able to enter into the lives of people as they create and reproduce culture in ways that quantitative methods alone cannot do.

Participant observation and field research remove the distance between sociologist and society and give life and continuity to Weber's notion of 'verstehen' as the basis for scholarly social understanding.   Schwartz's (1987) overt and eloquent defense of the unavoidable personal/subjective elements of ethnography, and the ways in which such research can be validated inter-subjectively, are well worth the reading. (See pages 10-13 and the Appendix).

Family sociology has also had its ethnographic classics, such as Komarovsky's *Blue Collar Marriage* (1962).   Using 58 families, she was able to do so deep and considered a study as to have provided a rich array of insights and testable hypotheses for the field.   We are convinced not by the quantity of her subjects, but by the inclusion of so many aspects of family process, including anecdotal data, that we are in touch with her "Glenton" families, and can induce from those families the kinds of systemic and interactional differences that create what are called "class differences" in conjugal relational patterns.

CONCLUSIONS

It must also be noted that if we are going to find out about the process of change as it is just beginning to occur, if we are to be sensitive to the emergence of new groups and what they tell us of sources and effects of larger social changes, and if we are to be tuned in to institutional changes at the familial-interpersonal level, we must be open to small-scale, time-consuming studies from which concepts may emerge into grounded theory, with perhaps new frameworks for understanding imminent changes. Particularly with new phenomena, and with samples that large data bases cannot reveal, the small-scale qualitative study is an irreplaceable source of social history, of small-group/family and large-scale changes on site, while they are evolving. We must also be open to samples that may find us before we find them, and be ready to study them and to enable them to know themselves better through our study results.

This type of phenomenon, where the social changes find the sociologist, has been part of my own experience. A study which began as an examination of religious feminists in the Protestant denominations, led to connections with married Roman Catholic priests and their wives, and finally resulted in a two-part study of the formation of new families in the face of specific prohibitions, stigmatized premarital relationships, and current strengths and problems within the marriages thus formed (Marciano, 1986; 1987).

The number of respondents in these studies have been small; some potential respondents were reluctant to answer questions; some are so eager for change that their perception of the likelihood of change in the institutional church seemed to be self-delusional. With all these factors in the study, and the sensitivity of the respondents to any apparent judgment of anything they have done, the participant-observer balance must be carefully maintained. Researcher self-awareness becomes the key to that balance; a careful reading of the subcultural language and cues of these groups is always necessary, so that nothing proceeds on any basis except truthfulness to maintain rapport and trust.

The cost in time and energy to do a study such as this is undeniable; the rewards, however, are exhilarating. And for that, my final note is one of thanks to those who have encouraged and received such studies, the true mentors of the discipline such as Marvin Sussman.

The art of sociology, as with any art, is intrinsically rewarding; but the support of respected and generous colleagues is the best gift we can receive and exchange. If we are prepared to give more of that support to each other, who knows what sociological wonders in the form of exciting and insightful studies we will all receive in return? ▓

REFERENCES

Bakanic, V., McPhail, C., & Simon, R.J. The manuscript review and decision-making process. *American Sociological Review*, 1987, 52,631-642.

Bernstein, R.   Critic of academe fears catch-22 in his success. *The New York Times*, December 28, 1877, sec. A, p.16.

Bowen, M.   Theory in the practice of psychotherapy. In P.J. Guerin (Ed.), *Family Therapy*. New York: Gardner Press, 1976.

Carden, M.L.   *Oneida: Utopian community to modern corporation*. New York: Harper Torchbooks, 1969.

Clinard, M.B.   The sociologist's quest for respectability.   In W.J. Filstead (Ed.), *Qualitative methodology*. Chicago: Rand McNally/Markham, 1970.

Coser, L.A.   Presidential address: Two methods in search of a substance. *American Sociological Review*, 1975, 40, 691-700.

Kafka, F.   *The Metamorphosis*. Translated and edited by S. Corngold. New York: Bantam Edition, 1972.

Komarovsky, M.   *Blue collar marriage*. New York: Knopf, 1962.

Grant, L., Ward, K.B., & Rong, X.L.   Is there an association between gender and methods in sociological research? *American Sociological Review*, 1987, 52, 856-862.

Marciano, T.D.   Married Roman Catholic priests: More 'married' than 'priests'?   Paper presented at NCFR Annual Meetings. Dearborn, MI., 1986.

Marciano, T.D.   From traditionalists to feminists: Wives of Roman Catholic priests.   Paper presented at NCFR Annual Meetings. Atlanta, GA., 1987.

Robertson, I.   *Sociology* (3rd ed.)   New York: Worth Pub., 1987.

Rosenberg, R.   *Beyond separate spheres: Intellectual roots of modern feminism*. New Haven: Yale University Press, 1982.

Schwartz, G.   *Beyond conformity or rebellion: Youth and authority in America*. Chicago: University of Chicago Press, 1987.

Sussman, M.B.   The social problems of the sociologist. *Social Problems*, 1964, 11, 215-225.

Sussman, M.B.   The urban kin network in the formulation of family theory. In R. Hill & R. Konig (Eds.), *Families east and west: Socialization and process and kinship ties*. Paris: Mouton & Co., 1970.

Whyte, W.F.   *Street Corner Society* (2nd ed.) Chicago: University of Chicago Press, 1966.

▒ 4 ▒

FAMILY AS A MANY SPLENDORED CONCEPT

Barbara H. Settles

INTRODUCTION

The purpose of this chapter is to examine ideas about families in the future, from the viewpoint both of what is likely and of what may be possible. Usually considerable energy is expended in defining the family as a term in theoretical discussions of the family. However, in this chapter, six approaches to the family as a concept are summarized: the family as an ideological abstraction, a romantic image, a unit of treatment, a last resort, a process, and a network. Although these topics do not exhaust the question of definition, they are suggestive of the range of ideas available. The difficulties of assessing the potential for change are discussed. The future itself, as an environmental setting for families, is examined to the limits of our imagination, invention, and innovation.

THE FAMILY AS AN IDEOLOGICAL ABSTRACTION

The family has been defined by scholars of many persuasions: law, religion, social science, medicine, and therapy. These definitions are used to label, diagnose, treat, harass, reward, and separate spheres of influence. Definitions that fit theories of political and religious order are developed for different purposes, for example, to discover the death of the family (Cooper, 1970) or to posit a profamily political position or to oppose divorce on theological grounds. The "family" as a concept is useful in strategic political planning because it has no common empirical referent. If you refer to the government, most listeners picture the same real government in their society, but with the idea of family, most

45

people picture either their own real family or the advertising world's stereotype of a family. Consequently, the content of the conclusions about a course of action and its impact on "the family" can vary greatly and yet be supported by the same arguments.

If one sees the past as a stable framework for family members, a classical view is developed of the family. This view or myth posits a standard for comparing today's families and for suggesting either a return to virtue or the total demise of the family in the future. Even popular magazines are now reexamining the myths about family (Gottschalk, 1977; *U.S. News and World Report*, 1980). Lantz, Schultz, and O'Hara (1977) called attention to this view as a refrain about "*families not being what they once were*" and noted examples from mid-nineteenth-century magazines.

The use of familial terminology in the structure of the church and in actual doctrine about familial regulation in religious organizations indicates the functionality of sacred definitions of family order. Although the specific content of the prescriptions and mandates varies from one religious persuasion to another, the essential element that the family is sacred, not secular, and thus the business of religion, has continued into the present.

The grouping of issues, which have been identified as profamily by conservative religious groups in America, is particularly interesting because of the basic conflicts among modernization, technological innovations, and family policy. Anti-abortion, antisecular sex education, anti-family planning, anti-Equal Rights Amendment, and anti-genetic research movements have been brought together under the rubric of being positive for the family as an institution. The uniting of the Fundamentalist Protestants, the Catholics, and the Orthodox Jews in America around these concerns has been a spectacular strategic success. Whether the coalition can hold together in the near future is more questionable. The unifying theme is the rejection of newer family forms and the availability of choice about family matters (Schulz, 1981). The fear of choice and change is often extremely high when innovations are introduced in any culture. However, the benefits that often occur to early adopters of innovation can undermine the penalties that traditionalists seek to impose.

Some prohibition-style regulation and enforcement are still likely in the immediate future as the profamily organizations influence legislative processes and bureaucratic policy enforcement. It would not be surprising to find abortion becoming a very scarce and expensive resource because of policy shifts (Ebaugh, Fuchs, & Haney, 1980; Issacson, 1981) and to find sex education remaining an undeveloped component of the school curriculum in the next decade (Brown, 1981). However, it is much less likely that the total package of conservative family policies will, in fact, take hold in ordinary practice in family life.

Other social forces make impractical the rigid and traditional sex roles and marital interaction (Bernard, 1972; Giele, 1979). The real likelihood of single parenthood and survivorship in old age makes it risky for women to be untrained and to remain outside the labor market. People cannot easily afford to invest for extended periods of time in social forms that are not compatible with the demands of daily life. For example, it is clear that Catholic couples in this society have adopted contraceptive practices not supported by religious doctrine and have worked out rationalizations to handle their conflicting beliefs and actions. On other issues, the society may have a short era of public policy on the family that is counter to public practice and that requires the service of an underworld to maintain current practice. Certainly, we have precedents in this century for prohibition strategies, and we know some of the costs and benefits.

THE FAMILY AS A ROMANTIC IMAGE

In contrast to the politicizing of a religious definition of the family, the romanticizing of the family found in the media and the peer group makes an impact not by forcing issues, but by structuring thought so that only certain choices are apparently good. Feminist scholars have studied intensely the way in which images of the family--and of men and women--in the media define our sense of self-worth (Elshtain, 1982). In the media, the woman who fails the "Ring around the Collar" or "Snoopy Sniffer" tests of good housekeeping is led to believe that her family is in grave peril. The man whose charcoal or mower won't start loses his manly leadership role in the family. These images and ideals of family competence do evoke laughter and ridicule, but they sell products and direct behavior.

The picture of the composition, the age, and the sex of family members has evolved in recent times. It is no longer so common to find the definitely older husband, the younger wife with the adolescent son, the elementary-school-aged daughter, and the baby as the ideal family. But other images linger on. Public opinion polls suggest that families have changed in how they view children's sex roles, but they still expect mothers to do the housework and fathers to be the breadwinners (Yankelovich, Skelly, & White, Inc., 1976-1977). For example, the husband's attitudes toward household work have changed from cohort to cohort (Wheeler & Avery, 1981), although housework is still not equitably shared in dual employment homes (Giele, 1979; Walker & Woods, 1976). Cherlin and Walters (1981) saw steadily increasing support for nontraditional sex roles over the 1970s. Identifying one's own family in the midst of these images leads to some evaluation of the quality of life and the success to be attributed to the family. If your experience comes close to these images and ideals, then the family situation is declared to be successful.

Revision in images can cause a previously positively evaluated situation to be rejected or vice versa. When behavior changes, the image or the family has to change also. In the twentieth century, this dictum has often meant that the specific family had to go: separating, divorcing, and remarrying (Bane, 1976). The process has been fixing the situation by changing who the players were rather than changing what the people themselves did. Some observers do find second marriages to have less similar marital partners than first marriages (Dean & Gurak, 1978), but in many cases, the new arrangement may simply repeat the problems of the old. The continuing future of a romantic view of a family is likely to continue to be emphasized by advertising and individuals because it simplifies the complexity of life for the moment.

THE FAMILY AS A UNIT OF TREATMENT

The growth of therapeutic intervention and family support systems demonstrates the possibility of the alteration of the internal family. Although the data are not complete on the efficacy of family therapy, the emotional and economic response has been positive. Families are willing to pay for intervention, and professionals are flocking to train as therapists (*Newsweek*, 1978).

The movement away from ordinary educational and media approaches to family intervention and toward peer counseling and family therapy is pronounced in professional circles. Peer counseling has the virtue of apparent economy because the peers volunteer their time. Supervision by professional family-life personnel may be quite minimal, and the support groups that have developed around every stress of family or public life are run on the commitment of the people involved. Marriage enrichment has developed as an approach to giving a similar experience to those who have no presenting problem (Mace & Mace, 1975).

Family therapy is far more expensive than peer counseling and seems to have a "severe" problem orientation. Families come into therapy after exhausting the resources of other areas of support, or by direct referral from these support systems. Commonly, a member of the family has been acting out in some labeled deviance that serves as the initial problem, for example, alcoholism, anorexia nervosa, juvenile delinquency, suicide attempts, drug abuse, school failure, job loss, mental or physical illness, or child or spouse abuse (Malcolm, 1978). The various approaches to therapeutic intervention are based on differing views of how families maintain interaction and structure within society and how the presenting problem is related to this family process. In the future, it may be possible to evaluate these approaches for relative quality, and families may be better advised about how to select supportive therapeutic intervention. In some cases, as the presenting problem itself is better understood, the family's role in the treatment or the alleviation of the problem may be reduced.

Family problems may be redefined as technological change alters how families relate to these problems. Corrective surgery reduced the need for the family of a child with a cleft palate to deal with stigma. As asthma and allergies have become better understood and the symptoms have become treatable, the family has been relieved of assuming the major responsibility for the allergic member's attacks. Schizophrenia, which has been seen as a family anomaly or biological fault, although not as well understood, is being reexamined for the scope and substance of the family's contribution (Falloon, Boyd, McGill, Razoni, Mosi, & Gilderman, 1982; Sheehan, 1982). The family may become less a unit of treatment in many situations when symptomatology becomes better understood. However, helping the family to adjust to the medical therapies will continue to be necessary even if the problems are redefined.

THE FAMILY AS A LAST RESORT

The family has often been viewed as the resource of last resort, as Robert Frost's poem (1946) suggested, "Home is the place where, when you have to go there, They have to take you in." The literature has long suggested that families, to the extent that they can manage the stress and continue as family units, provide for continuity in the care and treatment of the member with special needs (Hill & Hansen, 1964). The literature on singleness also notes that a person with few family ties may become lost in the institutional settings that replace the family support; hence, he or she may remain institutionalized long after the need has passed because no advocate asks questions. Stein (1981) suggested that the crucial issue may not be marriage versus singlehood but the strength of the support network.

The availability of family members and the concern of extended kin are crucial to the life course of elderly people. Shanas and Sussman (1981) found that old people without immediate kin are more likely than their contemporaries to be institutionalized. They further suggested that the twenty-first century will be characterized by the elderly "in search of a relative" to look after them or to take appropriate action on their behalf.

When a major institution such as the economy fails to function positively for individuals, they turn to family even in younger adult life. One of the rationales for keeping the large family home that was achieved when children were young is now to shelter adults and grandchildren in times of crisis.

Current conservative political groups suggest that the family might be the first resort for the solution of problems if other agencies and government do not intervene. The civil defense and disaster research has suggested that families use their family ties as emergency resources before turning to emergency programs (Drabek & Boggs, 1968).

THE FAMILY AS A PROCESS

Intimacy and continuity are basic needs that have been served by family units. When successful, the basic marital dyad has the capacity for enlarging the individual's experience in trust that supports both intimacy and a sense of continuity. In rapidly changing societies, shared meaning is a precious commodity. The game is far more often that of guessing if one understands another's gestures, words, and linguistic structure. The easy informality of the salesperson's use of list names or of the party introductions that identify people by name and job gives the appearance of closeness without the substance.

No one will claim that all or even a majority of families provide the meaningful relationships so widely discussed as a goal in the 1960s and 1970s. Families are strategic hunting grounds for such potential relationships. In contrast to work, school, and leisure settings, where temporary relationships and direct exchange are built into the institutions, families provide an opportunity for stability and indirect exchange as normative options. The fragility of the marital bond documented by divorce and separation seems not to be so closely paralleled in the parental role (Ryder, 1974). Although there has been some weakening of the parent's exclusive control, elasticity and continuity seem more typical of parents' and children's relationships. Adults today complain not of the empty nest, but of the returning "adult" child with his or her children in tow. Now that adoption has lost much of its stigma to the adoptee and the adopting parents, birth parents are seeking to contact the child, or the young adult seeks to open his or her records to find those parents (Cole, 1976). Often, foster parents care for children for longer than standards of practice recommend because the birth parents continue to make some efforts to continue duties.

THE FAMILY IN NETWORKS

The individually mobile nuclear family may attenuate extended kin ties during periods of stress in the movement between social positions. However, even these groups often reconstitute an interaction at life passage moments such as weddings, funerals, and hospitalizations. In addition to the kin of Western nostalgia, families today create their own relatives as needed. These friends who are called on to fill in for missing or nonfunctioning kin add substance to the potentiality for continuity, intimacy, and shared meaning as they are recruited for their commitments rather than by accident birth. The open family, as Constantine (1977) proposed it, may be quite purposeful in seeking such additional interactions.

Divorce and remarriage are adding great complexity to the potential for kin and family support systems (Kent, 1980). The child who is grandchild to a dozen other adults may surely find one or two suitable "soul affiliates." The price paid in loss of continuity may be regained in the intimacy and specialty of such relationships. The boundaries of these reconstituted families are more permeable, and a definition of new roles must be achieved (Walker & Messinger, 1979). The family chronicler who can keep the story straight and tell the great-grandchildren how the blended family all came about will have an important function.

In addition to the rearrangements of marriages and nuclear family units, improved health and life expectancy have added to the complexity of arranging each family vis-a-vis kin and friendship support systems. The family of four and five generations may become more frequent in spite of trends toward later marriage and childbearing (Glick, 1977; Reed, 1982). When elders are "deserted" by their children, the real situation is not that of young adults neglecting their parents but of already old "children" of 60 or 70 being unable to care for a 90-year-old parent and themselves as well (Steinmetz, 1981). There seems to be no reason not to expect this expansion of the population of the frail elderly to continue (Uhlenberg, 1980). The opportunity to recycle earlier friendship relationships is found because of these demographic trends. The widow may find her high-school sweetheart. The divorced may remarry. Friends may live together, pooling their pensions. Although new friendships and marriages are made possible by the extension of both life and good health, the actual prevalence depends on many other opportunity factors.

In the future, although family support may well increase, families are not likely to be designed around the concept of dependency. Financial and emotional dependency was fundamental to many of the restrictive norms on family dissolution and was sanctioned and supported the authority patterns within families. Continued dependency does not appear to be as essential as was previously thought, and it may be contrary to the demands of loving relationships (Saflios-Rothchild, 1977). The wave of married women with young children returning to the labor force has challenged the assumption of dependency by wives and has worried those desiring to continue patriarchal authority. Certainly, the many divorced women who support their children with little or no help from the father are not impressed by the dependency arrangements within nuclear families that ensure continued support (Schorr & Moen, 1979).

Children may also be "liberated" both from dependency and

from parental sponsorship for access to status and position. Currently, only a few children--those who have trust funds or social security benefits, those who are in foster care, and those who have their own jobs--are free of total reliance on family functioning for lifestyle. Child-rearing payments or family allowances are considered frequently in discussing family policy, but to date, they have met with little organized support (Kamerman & Kahn, 1978).

IMPLICATIONS OF FAMILY DEFINITIONS

These six ways of defining the family project different outcomes in futurist speculations:

Ideological Abstraction. These definitions appear to have considerable power for use by social movements. Currently, the traditional definitions are being used as rationales by the profamily coalitions in America and by religious groups throughout the world to promote more restrictive public policy on the family. A more clearly articulated statement of egalitarian family organization might have a similar utility for the women's movement.

Romantic Image. Romanticizing families continues to have great impact, especially in the public media. Although this view is generally rejected in family and women's studies literature and teaching, the fact that it must be countered as an approach suggests a tenacity in the concept. Today, there is an attempt to incorporate working wives and homemaking husbands into advertising. No doubt, the romantic image can be stretched to accommodate change without indicating the complexities underlying the change.

Unit of Treatment. This approach to refining an understanding of the family is the "cornerstone" of the conceptualization of family. The professions related to family therapy and family support services are still growing. The one caution is that some "family" problems may later be understood more fundamentally as medical, individual, or social problems.

Last Resort. Both the conservative political groups and the more liberal professional social-welfare service groups agree that this residual role is critical in understanding individuals and their resources. There is controversy over how to exploit the potentiality of the family to serve as last resort, but there is not much pressure, even from the political left, to replace the family in this function.

Process. Attention to definitions based on process have become important both for research and for therapeutic practice.

Since Burgess and Locke (1945) presented the rationale of a contemporary "modern" family based on companionship, the interest has shifted to analyzing how such a family may operate.

Networks. Both kin and friends as both interactive and integral parts of the family constellation are receiving scholarly attention and public policy notice. The complexity and the individualization of these networks appear to be increasing, but dependency as an organizing principal is declining.

CONCLUSION

These six approaches to the definition of the family illustrate a vast literature that refines, defines, argues, posits, and develops what the family may or may not be. If cross-cultural comparisons are included, there is a geometric progression of the amount of consideration of the idea of family. The fluid quality of the word *family* makes it especially useful for political propagandizing and as a residual variable in societal studies. In the future, it is not likely that the job of specifying what is meant by *family* will progress rapidly. Consensus would be costly for politicians and improbable for scholars. ▓

This chapter is an excerpt from "A Perspective on Tomorrow's Families," in M.B. Sussman & S.K. Steinmetz (Eds.), *Handbook of Marriage and the Family,* New York: Plenum Press, 1987. Permission to reprint this selection was granted by Plenum Press.

REFERENCES

Bane, M.J. *Here to stay: American families in the twentieth century.* New York: Basic Books, 1976.

Bernard, J. *The future of marriage.* New York: World Publishing, 1972.

Brown, L. (Ed.). *Sex education in the eighties: The challenge of healthy sexual revolution.* New York: Plenum Press, 1981.

Burgess, E.W., & Locke, H.J. *The family: From institution to companionship.* New York: American Book, 1945.

Cherlin, A., & Walters, P.B. Trends in United States men's and women's sex role attitudes 1972-78. *American Sociological Review,* 1981, 46(4), 453-460.

Cole, E. *Trends in family formation and dissolution: The case of adoptions, Implications for Policy.* Groves Conference, Kansas City, Missouri, March, 1976.

Constantine, L.L.   Open family: A life style for kids and other people. *The Family Coordinator*, 1977, 113-131.

Cooper, D.   *The death of the family.* New York: Vintage Books, 1970.

Dean, G., & Gurak, D.T.   Marital homogamy, the second time around. *Journal of Marriage and the Family*, 1978, 40 (3) 559-569.

Drabek, T.E., & Boggs, K.S.   Families in disaster: Reactions and relatives. *Journal of Marriage and the Family*, 1968, 30, 443-51.

Elshtain, J.B.   *The family in political thought.*   Amherst: University of Massachusetts Press, 1982.

Falloon, I.R.H., Boyd, J.L., Mc Gill, C.W., Razoni, J., Mosi, H.B., & Gilderman, A.M.   Family management in the prevention of the exacerbations of schizophrenia: A controlled study.   *The New England Journal of Medicine*, 1982, 306 (24), 1437-1440.

Frost, R.   *The poems of Robert Frost.* New York: Random House, Modern Library, 1946.

Giele, J.Z.   Changing sex roles and family structure. *Social Policy*, 1979, 9(4), 32-44.

Glick, P.C.   A demographer looks at American families. *Journal of Marriage and the Family*, 1975, 37, 15-26.

Gottschalk, E.C., Jr.   Exploring the myths about the American family. *Family Circle*, December 13, 1977.

Hill, R., & Hansen, D.A.   Families under stress.   In H. Christensen (Ed.),   *Handbook on marriage and the family.*   Chicago: Rand McNally, 1964.

Issacson, W. The battle over abortion. *Time*, April 6, 1981, 20-24.

Kamerman, S.B., & Kahn, A.J. (Ed.),   *Family policy, government, and families in fourteen countries.*   New York: Columbia University Press, 1978.

Kent, O.   Remarriage: A family systems perspective. *Social Casework*, 1980, 146-154.

Mace, D., & Mace, V.   Marriage and family enrichment: A new field. *The Family Coordinator*, 1975, 24, 171-173.

Malcolm, J.   A reporter at large: Family therapy, the one way mirror. *The New Yorker*, May 15, 1978, 39.

Reed, J.D.   The new baby boom. *Time*, February 22, 1982, 52-58.

Ryder, N.B.   The family in developed countries. *Scientific American*, 1974, 231(3), 122-132.

Safilios-Rothschild, C. *Love, sex, and sex roles.* Englewood Cliffs, NJ: Prentice Hall, 1977.

Schorr, A., & Moen, P.   The single parent and public policy. *Social Policy*, 1979, March-April), 15-20.

Schulz, D.A.   *Speculation on the future of the family: Before bearing.*   Groves Conference on Marriage and the Family,   Mt. Pocono, PA., May 1981.

Shanas, E., & Sussman, M.B.   *Aging: Stability and change in the family.* New York: Academic Press, 1981.

Stein, P.J. *Singlehood,* Presented at Groves Conference on Marriage and the Family, Mt. Pocono, PA., May 1981.
Steinmetz, S.K. Elder care and the middle aged family. Presentation at the American Home Economics Association, Atlantic City, N.J., June, 1981.
Steinmetz, S.K. Elder Abuse. *Aging,* 1981, 315-316, (Jan/Feb), 6-10.
U.S. News and World Report, 1980.
Uhlenberg, P. Death and the family. *Journal of Family History,* 1980, 15(3), 313-320.
Walker, K.N., & Messinger, L. Remarriage after divorce: Dissolution and reconstruction of family boundaries. *Family Process,* 1979, 18, 185-191.
Walker, K.E., & Woods, M.E. *Time use: A measure of household production of family goods and services.* Washington, D.C., American Home Economics Association, 1976.
Wheeler, C., & Avery, R.D. Division of household labor in the family. *Home Economics Research Journal,* 1981, 10 (1), 10-20.
Yankelovich, Skelly, & White, Inc. *Raising children in a changing society.* The General Mills Report, 1976-1977.

▓ 5 ▓

CROSS-CULTURAL RESEARCH: A SAFARI APPROACH

Linda K. Matocha

INTRODUCTION

The study of family is enlightening to researchers of social behavior, but it can be equally frustrating as well. It seems that the "true" essence of what is meant by family eludes even the most diligent scholar. When there is an added dimension of studying families of other cultures, the process appears to become impossible. With this added complication, why then do scholars need to study families of other cultures?

Historically, the family was studied as a small social system based on biological and sociological heritages (Queen & Adams, 1952). Contemporary family scientists recognize that the cross-cultural study of family makes it possible to appreciate similarities and differences between and among cultures; be more sensitive to cultural heritages; and enlighten others to possible relationships about families with variables not previously considered (Cogswell & Sussman, 1972; Nye & Berardo, 1973). Unfortunately, many researchers tend to be ethnocentric in their study of the family, thus limiting their knowledge acquisition by not allowing a more objective evaluation of families representing different cultures and different forms (Brown, 1981; Nye & Berardo, 1973; Queen & Habenstein, 1974; Sussman, 1974). The study of other cultures brings the study of our own culture into perspective and allows for a more in-depth understanding of American family life by providing solutions that are not necessarily evident in studying the family from a narrow perspective of a single culture. As Sussman (1965:27) has noted:

> Cross-cultural studies are necessary to determine the bases for harmonizing apparently incompatible and contradictory bureaucratic and familial norms.

57

METHODOLOGICAL ISSUES

Studying the cross-cultural variations of families does pose some methodological problems. The longitudinal ethnographic or phenomenological method is probably the most preferred because it enables researchers to more fully understand family processes in the family system and how the family system interacts with its sociological environment. The longitudinal method has been used for studying the American family (Burgess & Wallin, 1953; Hill, 1970b; Peters, 1984), and would be a valuable tool for studying other cultures. However, this methodology has major difficulties associated with it. First, the life cycle of the family covers an extended time frame, and may exceed the researcher's life expectancy. Second, it is difficult to locate families willing to participate in research encompassing such a long period of time and the attrition rate leaves the study vulnerable to error. Third, families who agree to participate may become "trained" by the research team to respond in a certain manner. A fourth difficulty results from problems in changing the design of the study even though families themselves change over time (Sussman, 1966). Finally, the expense (in time and money) of studying families over such a long period of time is frequently too high. Funding agencies are unwilling to allocate the necessary funds to a single study covering an extended period of time, and researchers find the risk too high if they put all their academic "eggs" in a single basket.

Alternative innovative methodologies (retrospective, cross-sectional, intergenerational and segmented longitudinal) have been developed by researchers to study families, but each have major difficulties (Hill, 1970a). Although the difficulties are known to family researchers, each methodology enables the addition of different kinds of knowledge to further the understanding of family. Sussman (1958:36) recognized the need for innovative methodologies when he stated:

> Every once in a while, like a breath of spring air, some one comes up with a research procedure that stirs the imagination and challenges the doctrinaire 'ask a question' approach. Though often crude in design and more frequently never replicated, these nuggets of imaginative research methodology deserve a hearing.

The Safari Approach

An additional methodology exists that satisfies this concept and needs to be examined and developed. This approach can add needed insights into family and yet do not constitute prohibitive spending of extra time and money in today's world of economic restrictions. Researchers of families frequently travel to different countries in various capacities. Can research be done on these brief, often very busy trips? This answer is emphatically *"yes"* because research is *"the systematic investigation into a subject in order to discover new knowledge, principles, and facts"* (Sussman, 1964). Research done on "safari" can add new knowledge and facts about another culture. Thus the systematic collection of data and utilization of appropriate methodologies, e.g., content analysis, secondary analysis of existing data, and historical/comparative analysis are critical for assuring quality research (Babbie, 1983). Webb and others (1966) have labeled these methodologies "unobtrusive research." They believed that you can study individuals by observing closely "what they leave behind." Researchers are trained in closely observing those variables which they are studying. These observational skills are useful for conducting "safari research"--research into new cultural territories on a short-term basis. What is needed, however, is a systematic investigation resulting in a methodology that family scientists can utilize.

Such a systematic investigation was attempted with a two week "safari" to Australia and New Zealand. The data gathering was a three stage process:

Stage 1: Preliminary gathering of knowledge and equipment. Any well prepared hunter going on safari learns as much as possible about the country, the quarry, and the rules of the hunt. The hunt in this instance was to promote friendly relations with Australians and New Zealanders through the medium of scientific, professional and technical exchange in the People to People Citizen Ambassador Program. Delegation members were given professional objectives with the overall objective of reaching out in friendship to the people of other nations to promote world peace.

The quarry was knowledge about the people and families of the two countries, and how transcultural concepts were integrated into nursing. Knowledge about the country included the history and geography of the country, the type of government, and pertinent demographics. The demographics gathered included population numbers with annual growth rate, religious affiliation, educational preparation of the people, budget, state of the economy with inflation rate and gross national product, cultural origins of the people, and specific demographics concerning the health of the people. Health demographics that seemed to give the most information about families and health included infant mortality, life expectancy, incidence of disease and illness specific to males and females, expenditures of health services, and the incidence of health problems of the different culture groups.

A review of journals, newspapers and magazines from the country gave additional insight into how the country's journalists saw the problems and strengths of life in their respective countries. Speaking to others who have either lived in or visited the country added a personal perspective of what to expect in the daily life of a visitor. Discussing music, food, acceptable dress in various atmospheres and environments, available leisure activities, and even where and how to shop provides the hunter with knowledge of family life in another culture.

Physical equipment for the "safari" is really quite simple. The essentials are a good camera, tape recorder and an extra suitcase to carry back trophies. A diary should be kept with entries denoting where the information can be found and in what context it was gathered. The traveler finds that it will be referred to frequently for information and direction for further observations.

Stage II: On safari. The hunt continues into the country. The "hunter" researcher continues observing but aims at a specific quarry. Even though much of the researcher's time is spent in meetings or in the hotel room, there are pertinent observations which can be made. Local newspapers, telephone books, television programming, bulletin boards, and pamphlets add enlightening data about the family and life in another culture.

Each person that the hunter meets acts as a guide to the elusive quarry of understanding family life in another culture. In Australia and New Zealand there were informal meetings (tea time) between the scheduled more formal meetings. At this time information was gathered on many levels from the general perceptions of the informants on what was occurring in family life to the more individual perceptions of the informant's own family life. Informants may even be encouraged to share family pictures with the researcher. The researcher should note who is in the pictures and what they are doing.

Wherever the researcher goes, information can be gathered. On the trip to Australia and New Zealand, much time was spent in buses with the same bus driver. Many drivers were willing and interested in providing information about their culture and family life. If time allows, the researcher should go where the locals go during their everyday lives--to the food market, to visit homes located in different areas, to the zoo, to shopping areas. Each can provide insight, and have the added bonus of being very enjoyable. Audio tapes of contemporary music of the country can easily be purchased. The researcher should document the type of music preferred by different groups. The art of the country should also be documented. In Australia and New Zealand the use of color is much more vivid and used with little shadings, as compared to the U.S. Reds, yellows, and vivid earth tones are used with black to emphasize life. The humor of the people should also be documented. Note what makes people laugh, what jokes are told, and the context in which they are told. The diary of the researcher becomes a repository of information of not only what was said but also of the context in which it was said and gathered.

Pictures add to the documentation. Take pictures of things

that interest the researcher and that record the everyday lives of the people. Pictures of as many of the entries in the diary as possible are important. They tend to capture the feelings of the moment so that they can be more closely examined later. Include pictures that capture people and their family lives across the life span. Pictures may include hospitals or places of birth; they may include schools, universities, and places of work; and finally they may include resting places of the dead. Careful documentation of when and where the pictures were taken is very important; do not think that you will remember later.

Stage III: Mounting the trophies.   After the trip has been completed, pictures developed, and time is available, reread the diary of the trip. Determine how best to collate the data. Different arrangements of the data may provide the researcher with different perspectives of the people and family life. Many questions may be answered, but many more will arise. Further study is very important, but the knowledge gained from even a short safari can be shared with others who are interested. The pictures, demographics, and verbal stories of what occurred during the trip can assist others in gaining insight. The music can be shared during the presentation if it adds to the "flavor." The researcher, like the hunter, will find that others will have many questions about the quarry, the people and family life. That is the function of research--to discover knowledge, add to knowledge, and encourage the thirst of knowledge in others.

CONCLUSION

This is the framework of a type of research methodology that researchers need to use more to their advantage. It is a flexible, yet straightforward methodology that enables one to gain insights into families and cultures. This insight can then be applied within the context and interpretations of future research and exchange of knowledge. Each researcher brings to this research process areas of expertise that could be used to further refine this methodology to meet the needs of the moment. ▓

REFERENCES

Babbie, E.   *The practice of social research* (3rd ed.). Belmont, California: Wadsworth Publishing Company, 1983.
Brown, M.S.   Culture and childbearing.   In A.L. Clark (Ed.), *Culture and childbearing.*   Philadelphia: F.A. Davis, Company, 1981.

Burgess, E., & Wallin, P. *Engagement and marriage.* Philadelphia: J.B. Lippincott Company, 1953.

Cogswell, B.E., & Sussman, M.B. Advances in comparative family research. In M.B. Sussman & B.E. Cogswell (Eds.), *Cross-national family research.* Leiden, Netherlands: E.J. Brill, 1972.

Hill, R. Methodological issues in family development research. In N.W. Ackerman (Ed.), *Family process.* New York: Basic Books, Inc., 1970. (a)

Hill, R. The three generation research design method for studying family and social change. In R. Hill & R. Konig (Eds.), *Families in east and west.* Paris: Mouton & Company, 1970. (b)

Nye, F.A., & Berardo, F.M. *The family its structure and interaction.* New York: Macmillan Publishing Company, Inc., 1973.

Peters, J.F. Role socialization through the life cycle of the Yanomamo: The developmental approach to the study of family in a preliterate society. *Journal of Comparative Family Studies,* 1984, 15(2), 151-174.

Queen, S.A., & Adams, J.B. *The family in various cultures.* Chicago: J.B. Lippincott Company, 1952.

Queen, S.A., & Habenstein, R.W. *The family in various cultures* (4th ed.). Philadelphia: J.B. Lippincott Company, 1974.

Sussman, M.B. New approaches in family research: A symposium. *Marriage and Family Living,* 1958, 20(1), 36-38.

Sussman, M. B. Experimental research. In H.T. Christensen (Ed.), *Handbook of marriage and the family.* New York: Rand McNally, 1964.

Sussman, M.B. The urban kin network in the formulation of family theory. Paper presented at the Ninth International Seminar on Family Research, Tokyo, Japan, September, 1965.

Sussman, M.B. The measure of family measurement. In C. Chilman (Ed.), *Approaches to measurement of family change: Research reports #4.* Welfare Administration, DHEW. Washington, D.C.: U.S. Government Printing Office, August, 1966.

Sussman, M.B. Issues and developments in family sociology in the 1970s. In M.S. Archer (Ed.), *Current research in sociology.* (Proceeding of the VIIIth World Congress of Sociology, Toronto, Canada). Paris: Mouton & Company, 1974.

Webb, E. et al. *Unobtrusive measures: Nonreactive research in the social sciences.* Chicago: Rand McNally, 1966.

▓ 6 ▓

INTIMATE VS NON-INTIMATE SUPPORT NETWORKS

James W. Ramey

INTRODUCTION

What is a support group? This term, a recent addition to our lexicon, has been used rather loosely to define everything from the owners of 1964 T-Birds to residents of a halfway house for recent prisoners. Yet these two examples are obviously vastly different in terms of the degree of intimacy involved, the intensity of the relationships, and the importance of outcomes to the participants.

Family Support Systems

Family support systems, the ways in which family members provide mutually beneficial, reciprocal support, were investigated and brought to the attention of academics about three decades ago (Sussman, 1959). During the decades following World War II, when 20% of the population was on the move every year, Sussman was concerned about whether traditional family functions would continue to operate. Some of his early work dealt with parents supporting their grown children at a critical juncture such as needing help to buy a house or to start a business (Sussman & Burchinal, 1962). Later work examined intergenerational support through wills and trusts (Sussman, Cates, & Smith, 1970). His research on how the geographically widespread family nevertheless manages to support aging parents led to further research on ways to strengthen family support in order to minimize the need for institutionalization of the elderly parent (Sussman & Ramey, 1978). Governmental support appeared to be supplanting family support in many ways, and as always, when the government does it, it costs five times as much.

A key finding in all of these studies was that family still

means something special in our various U. S. cultures.   Despite problems introduced by distance, which results in family members being spread from coast to coast, parents continue to support children long after they have left the nest and children continue to support their elderly parents as long as it is physically possible to do so.   The intergenerational links are still strong, as demonstrated not only by wills and trusts but also by pre-nuptial agreements through which older people protect existing family rights (Sussman, 1983; Sussman & Jeter, 1985).

How do strong family ties develop?   The most obvious way, of course, is the sexual vector that cements various bonds among the group for varying lengths of time.   This includes marriages, and living together relationships, as well as separations, divorces and deaths. A derivative of the sexual vector is being a member of the family   by blood or marriage.   This used to be a fairly straight-forward relationship: children grew up with two parents, four grandparents, uncles, aunts, cousins, nieces and nephews.   Today families often involve one or more remarriages, so the family ties are tangled and some come untied.

Nevertheless, the other vectors along which these ties are expressed remain fairly constant.   Families are notoriously involved in complex emotional relationships, perhaps now more than ever, given the complexity of some contemporary families.   They are also pulled together or torn apart on the religious vector.   Many of us can remember being told by our parents to be polite to, but have nothing to do with, kin who belonged to a different religion.

Likewise, families are often involved in complex economic re-lationships, which again can lead to great intimacy or frosty avoid-ance.   Some families have made lasting marks on our society because of their complex intellectual, cultural or career involvement.   Final-ly, families share social intimacy to a greater or lesser degree.

But what about people without family ties?   Blood may be stronger than water, but when we read today about support groups the   writers are seldom telling us about family ties.   Instead they are touting Al-Anon, EST, a feminist group, a writer's group, owners of car phones, Mason's, PTA, The League of Women Voters, Gay pride, or any of hundreds of other causes.   Are all of these groups truly support groups?

Non-Traditional Support Groups

Support can be provided by groups that do not represent tra-ditional families.   Research on non-traditional families (Ramey, 1975) would suggest that members of group marriages clearly seemed to exhibit the degree of intimacy one would expect in a family. However, in other groups such as communes, the level of intimacy varies with each commune.   Swingers, on the other hand, were certainly sexually intimate but did not generally seem to exhibit anything like the degree of intimacy one finds in families, although there were exceptions. The one non-traditional life style that appeared to involve many of the types of intimacy found in extended families was the sexually open relationship network.

This kind of network seemed to grow without conscious effort from the linkages between couples and singles, straight, gays or bisexuals, who initially became sexually intimate friends. Soon they found themselves interacting with spouses and other friends of friends who also subscribed to the notion that openly allowing but not insisting on sexual intimacy among friends was more honest and rewarding than the usual clandestine meetings (Ramey, 1975). An unusually tight support network appeared to develop around such relationships.

These networks also involved social, economic, emotional, religious, intellectual, sexual, career, cultural, economic and/or family ties, with varying degrees of relationship/commitment on each of these vectors. As these individuals in sexually open relationships were fond of pointing out, their intimate networks were based on chosen relationships, rather than family networks tied to one's birth. Thus, they suggested, their networks were likely to be more enduring than family networks.

Over time, each relationship developed differing degrees of complexity based on the strength of the links on each vector. Thus A might be strongly linked to B on the emotional, social, intellectual, and sexual vectors but linked to B's husband only on religious, social and economic vectors. B, in turn might be linked to C only on sexual and emotional vectors but to C's wife on family, intellectual and social vectors.

While the nature of most of these vectors is self evident, it should be noted that on the family vector, people considered family sharing to include activities such as vacationing, renting a boat or RV, taking care of a friend's children, or simply doing family things together. Several people spoke of helping the friend's child get into a particular college. On the career vector, people mentioned such things as being a mentor for a younger friend coming up, showing a friend the ropes in a job, helping with a career or job problem or using the informal network to find a new position. On the economic or financial vector, friends went into business together, lent money or co-signed a note, paid for a friend's tuition or his/her child's tuition, sent a friend's child on a trip abroad or in other ways helped out financially.

Intimate networks. For the purposes of this research, an intimate network support system was defined as a network in which each and every member named every other member as a part of the group. Of course some other people were named as well, but they were not considered to be part of this particular support system unless the previous condition was met. Although not all of these people were sexually involved with one another, they all subscribed to the notion that sexual involvement with friends was an openly acceptable practice.

One might ask: Does such a sexually intimate network differ from any other support network of friends? Is sexual involvement the only difference? Does sexual involvement have any impact at all? Since a large number of both men and women report involvement in extramarital relations, there is a high probability that many of the friends in a non-intimate support group are actually sexually involved with one another.

What does intimate mean? People in sexually open intimate networks insist that intimate does not necessarily mean sexually intimate. They note that at any given time, perhaps a third of their intimate friends are not sexually involved outside their primary relationship, and many have never been so involved. If we accept this self-definition, which views intimacy not in sexual terms but in some other way, at face value then would the difference between intimate and non-intimate support networks be eliminated?

People in intimate networks insist that a non-intimate support network is a contradiction in terms, since intimacy and commitment are essential ingredients of a support network. They define intimacy as willingness to share vulnerability. Many will say something about the value of sexual openness as a short-cut to sharing vulnerability. These people point out that non-sexual group members usually take a long time developing sufficient trust to share vulnerability, especially in a co-ed group and particularly on the part of the males in the group.

They feel that once you've gone to bed with a person, you have made yourself so vulnerable that it is easy to share other kinds of vulnerability such as sharing secrets, insecurities, plans, personal deficiencies, and illegal or illicit actions or information.

These people have accepted sexual intimacy within the group as their mechanism for defining the group. In a sense they may also be suggesting that sharing sexual vulnerability is an earmark of the support network, i.e., it defines the network because it is the purpose around which the support network grew. It is important to realize that vulnerability can also be shared on other vectors and indeed knowing the vector on which a support group shares vulnerability may provide great insight into the focus of the group.

For example, a family might choose to make itself financially vulnerable with other family members, in order to support an aged parent, to help put a child through college, start a business or buy a home. A group of friends might decide to live together for the summer, making themselves mutually vulnerable to each other's personal habits and idiosyncrasies. A group might make itself emotionally vulnerable by undertaking group therapy. A group of Wall Street lawyers might make itself legally vulnerable by sharing insider information about a coming corporate merger or acquisition.

Maintaining Groupness

We can probably all agree that any support group, in order to maintain its sense of groupness, must be concerned about three things: the purpose of the group, the group membership, and the effectiveness of the group in achieving its purpose. There must be a clear definition of purpose, understood and agreed upon by all members of the group. For example, is the group a consciousness raising group for women, or is this a family group that is determined to keep Mama out of the nursing home, or is this a stock purchase group, or is this an Alcoholics Anonymous group? In each case not only is the goal or purpose clear, but it is unanimously subscribed to by all members.

Boundary maintenance can be reasonably defined in terms of all group members being able to readily list all members. If the boundary is so permeable that members are unsure who is a member and who is not then it is not likely that they will be willing to make themselves vulnerable because they will not trust the group to respect the confidentiality of their revelations.

Commitment to the support group can be translated as continued availability to other group members along at least some of the vectors that the group feels are pertinent to achieving its purpose, e.g., social, sexual, emotional, economic, intellectual, cultural, religious, family and career. This continued availability depends on each member's continuing belief that the group is on target and achieving or working towards its purpose or goal in an acceptable manner and at an reasonable pace.

A GROUP INTIMACY SCALE

The efficacy of self-defined groups has traditionally been measured by how long the group lasts. An instrument that could measure the degree to which a group could be considered a support group would be beneficial. This instrument would have considerable utility if it could rank groups not only on the degree to which they function as support groups, but also on their relative cohesiveness as seen by their members. Such a scale would measure the degree of commitment members have to the group as well as the degree to which they have shared vulnerability. The Group Intimacy Scale, described below, fills this void (See Figure 1).

Scoring

To arrive at a score, add up the "yes" answers to questions 10-23. Scores between 0 - 3 suggest that the group is NOT a support group; between 4 - 7 suggest that the group is a support group, and 8 or higher indicate that the group is a strong support group.

GROUP INTIMACY SCALE

Please answer the following questions about your group. The composite returns from all the members of your group will determine the degree to which your group is defined by its members as a support group.

1. How long has your group been in existence?
2. Name the members of your group.
3. What is the purpose of your group?
4. Is it making reasonable progress toward achieving this purpose?
5. Do you accept new members?
6. Must new members meet some sort of membership requirements?
7. How many new members have joined your group since it began?
8. How many new members have joined your group in the past year?
9. Why did members leave the group?
10. Is your financial contribution to your group substantial?
11. Are you making a substantial time commitment to your group?
12. Have you been involved in illegal activity with group members?
13. Have you used illegal drugs with group members?
14. Have you shared confidential information with your group?
15. Have you revealed personal secrets to your group?
16. Does your group share secret information known to no one else?
17. Do group members share and/or help deal with personal problems?
18. Do group members share/help with family/relationship problems?
19. Do group members help deal with/correct educational deficiencies?
20. Is sexual activity between group members openly acknowledged?
21. Does your group own real estate, securities or other assets?
22. Can members get group financial support in time of need?
23. Does your group provide career help to members?
24. In what ways are members similar: Age_____ Sex_____ Couples_____
    Avocation_____ Ethnicity_____ Nationality_____ Educational level_____
    Income_____ Religion_____ Career_____ Same Family_____ Singles_____
    Social Class_____ Sexual Preference_____ Other_____
25. What is the age range in your group? From _____ to _____.
26. Indicate the degree to which you are committed to or have made yourself vulnerable to your group on each of the following vectors using a scale of 1 (low) to 5 (high):

Vector	Commitment	Vulnerability	Combined Score
Social	____	____	____
Sexual	____	____	____
Career	____	____	____
Economic	____	____	____
Family	____	____	____
Religious	____	____	____
Cultural	____	____	____
Emotional	____	____	____
Intellectual	____	____	____

Figure 1. A Group Intimacy Scale.

Questions 25 and 26 enable us to know whether the group is a strong support group or indeed, a support group at all. (Question 24 provides SES data about the group.) Question 26 is scored by combining the commitment and vulnerability score for each vector in the column headed "Combined Score." These scores tell us more about the complexity of the particular group--whether it is a less complex group or a blend of relationships--and the strength of the relationships.

A combined score of 0-3 suggests little commitment or vulnerability, a score of 4-6 suggests average involvement, and a score of 7 or higher suggests strong involvement on that vector. A low vector score indicates either that the group is not very complex, not very old or both. A high score would suggest a very complex group with many interrelationships.

The total composite scores are interpreted in a similar manner. Scores between 0-31 suggest little intimacy and indicate that the group is probably not a support group. A composite score of 32 - 59 indicates average intimacy and support involvement, and a score of 60 or higher indicates a very strong support group. By juxtaposing the two scores derived from this scale we are able to compare the cumulative self-reported estimates of intimacy with cumulative group answers to specific questions that explicitly define intimacy.

Finally, question 25 provides critical information about the likely longevity of the group. Groups with a narrow age range tend to die with their founders. Not only is new blood essential to growth, but the new members must be younger members or the group will slowly gray and then die. It is easy to recognize problems of aging and the need for intergenerational support in families but not so easy in groups and organizations.

If we look around us, we recognize some groups, such as the VFW and the Buggy Whip Manufacturers Association as groups that have flourished in the past but are far past their prime today because they lack new blood. The Shakers were very strong as long as they took in orphans and raised them, but since they did not believe in marriage and sexual intercourse, few new members joined and the group died out.

Other strong support groups have lost their original purpose, developed permeable boundaries, or changed the nature of the group so that it is no longer a support group. The Amana Colonies and the Oneida Community are both examples in which the shift was from an intimate support group to an economic group operating through corporations.

Testing the Group Intimacy Scale

In order to test the ability of the Group Intimacy Scale to identify the nature and complexity of groups, the scale was administered to two groups, a men's support group and a women's consciousness-raising group.

Men's support group. The first was a 10 member men's support group that has existed for less than a year. All members of this group were able to name all other members. The purpose of this group is to explore ways in which men are less than equal with women, to share feelings of inadequacy, and to build emotional support. The group members all feel that they are moving in this direction at a reasonable pace.

This group has not thought about accepting new members. They are concerned about maximizing trust and their ability to work together. They do not feel that they are ready to take in new members. They recognize that this would reconstitute the group as a new and different group, requiring them to rebuild trust before they would be willing to share vulnerability. All the current members are charter members.

Group members feel that they are making a significant time commitment, meeting for several hours one night a week despite busy schedules. They are not making a big financial commitment. The men have not shared illegal activities, drugs or sexual activity. However, members have shared personal secrets and confidential information, discussed personal and family problems and provided limited career help.

Group members feel that the consensus they have developed so far about ways in which they feel men are "second class" citizens constitutes a form of secret information known only to the group. Group members have not helped with educational deficiencies, nor do they own joint property or other financial assets. They are not sure whether financial help might be forthcoming from the group in case of individual need.

On questions 10 - 23 this group has a score of 7, which suggests that the group is indeed a support group but not yet a strong support group. The age range is from 32 to 64, a reasonably broad range suggesting that if new members are eventually admitted to this group and it continues to strive toward its goal, it could conceivably outlive some of the current members. All are white, married, members of the same religious group, and born in the USA. The educational levels range from college graduate to Ph.D. The men are in professional, educational or entrepreneurial positions and they have higher than average incomes.

Composite scores on question 26 are Social: 7, Sexual: 0, Career: 2, Economic: 2, Family: 0, Religious: 8, Cultural: 4, Emotional: 8, and Intellectual: 7. Thus the group exhibits average involvement on the cultural vector and strong involvement on the social, religious, emotional and intellectual vectors. An overall score of 38 places this group in the low average complexity level, which is not surprising despite the nature of the group, when you consider that it is less than a year old.

Women's consciousness-raising group.   A contrasting group, also less than a year old, is a women's consciousness-raising group. All sixteen members are able to name all other members of the group, but fellowship, not the stated goal of consciousness-raising, was stated by some members as their reason for membership. All members thought they were making reasonable progress toward their respective goals.

The group is closed and does not expect to accept new members. It has lost two members since it began, both to competing commitments. Members feel they have made substantial time but not financial commitments to the group. There have been no illegal activities or drug or sexual sharing among group members. The group has shared confidential information and personal secrets have been revealed, but the group does not feel that it possesses any secret information known only to the group.

The group has helped with personal problems, but not to any great extent. The women do not share family or relationship problems or educational problems. They own no joint assets, are not available for individual financial support in time of need, and do not provide career help.

The group is comprised of married, single, and divorced women ranging in age from the 20s to the 60s, who are white, middle class, college graduates with a similar religious affiliation. All of the women are at least second-generation Americans.

On questions 10 - 23 this group has a score of 4, which barely suggests that the group is a support group. Composite scores on question 26 were Social: 7, Sexual: 0, Career: 0, Economic: 0, Family: 0, Religious: 8, Cultural: 6, Emotional: 4, and Intellectual: 4. This group exhibits high average involvement on the cultural vector and strong involvement on the social and religious vectors, but the overall score of 29 suggests that in spite of its stated purpose this group is not a support group. So far the members of this group have shown little commitment other than on the social and religious vectors and have made themselves vulnerable only in the area of sharing religious beliefs. It is interesting to note that there is minimal emotional and intellectual involvement in spite of the shared religious based and highly emotion-laden goal of the group. The clue to the problem probably lies in the reports of several of the participants that they were in the group for social reasons, since they felt that their consciousness had been raised years ago.

CONCLUSION

The Group Intimacy Scale was developed to provide a measure of the commitment and complexity of relationships among members in support groups.   The scale was administered to two groups: a men's support group and a women's consciousness-raising group. Although limited, the results between the two groups, which on the surface appear to have similar goals, reveal the ability of the Group Intimacy Scale to differentiate the intimacy level of groups. Further testing of this instrument, especially with "New Age," humanistic oriented groups such as The Possible Society, would be of interest.   One could expect that these groups would indeed be a support network with the high level of intimacy usually associated with family support networks. ▓

REFERENCES

Ramey, J.W.   Intimate groups and networks: Frequency of sexually open marriage. *Family Coordinator*, 1975, 24-4.

Ramey, J.W., & Sussman, M.B.   Incentives to promote home care of the aged,   9th World Congress on Sociology, Uppsala, Sweden, August, 1978.

Sussman, M.B.   The isolated nuclear family: Fact or fiction. *Social Problems*, 1959, 6(Spring), 33-340.

Sussman, M.B., & Burchinal, L.G.   Parental aid to married children: Implications for family functioning. *Marriage and Family Living*, 1962, 24(August), 231-240.

Sussman, M.B., Cates, J.N., & Smith, D.T. *The family and inheritance.* NY: Russell Sage Foundation, 1970.

Sussman, M.B.   (Burgess address) Law and legal systems: The family connection. *Journal of Marriage and the Family*, 1983, 45(Feb.), 9-21.

Sussman, M.B., & Jeter, K.   Each couple should develop a marriage contract.   In   H. Feldman & M. Feldman (Eds.), *Current controversies in marriage and family.* Beverly Hills, CA: Sage, 1985.

HEAD START'S INFLUENCE ON PARENTAL AND CHILD COMPETENCE

Donald L. Peters

INTRODUCTION

Recent reviews and syntheses of research on the effectiveness of Head Start continue to substantiate the potential short-term benefits of regular participation for low-income children, and to some degree for handicapped children, on some measures of cognitive and social development (McKey, Condelli, Ganson, Barrett, McConkey, & Plantz, 1985). Longitudinal analyses also document the intermediate and long-term effects of Head Start and similar early childhood intervention programs (Clement, Schweinhart, Barnett, Epstein, & Weikart, 1984; Lazar, Darlington, Murray, Royce, & Snipper, 1982). In the latter case, it appears that early intervention has a strong and continuing effect on children's ability to cope with the basic demands of schooling right through the completion of high school.

It also seems that the relationship between such early intervention and the production of long-term effects is not a simple one (Lazar *et al.*, 1982; Woodhead, 1985). Rather, the short-term effects of participation are mediated within a context of other variables in the home and school social environment, both during the period of intervention and throughout the later stages of education.

It is generally hypothesized that this complex relationship involves a combination of the child's developmental susceptibility to environmental input, quantitative variability in the amount of intervention offered, and the breadth of the effort expended to alter the child's context (MacDonald, 1986; Clement *et al.*, 1984; Lazar *et al.*, 1982; Woodhead, 1985). By creating a discontinuity between the child's prior, current and future environment, intervention can change the course of the child's development through the reduction

73

of risk factors in the environmental context and the creation of greater learning opportunities (Peters & Kontos, 1987). These interventions take into consideration the child-learning environment as well as parent-child relationships. How these relationships may be altered is determined, at least in part, by the purposes, nature, and guiding principles of the intervention programs planned.

FAMILY LEARNING ENVIRONMENTS

For all children there are inherent risks and opportunities in their family's physical and social environment (Garbarino, 1982). Opportunities exist when the family environment provides for adaptive, growth enhancing experiences at some optimal level for the child's current developmental status. Risks to development can result from direct threats, a lack of opportunities as well as from acts of commission by parents and other family members (Peters & Kontos, 1987). These risks include the obvious biological ones involving physical abuse, inadequate diet, unsanitary or disease-ridden living conditions, as well as situations that are more subtle and involve psychological damage or deprivation such as those involving father absence or non-normative family transitions (McCubbin, Sussman, & Patterson, 1983). Often they involve the interaction of several factors.

Assessment of the risks and opportunities for a particular child or group of children requires an understanding of the child's attributes as well as the salient features of the family context. Describing the family context and assessing its risks and opportunities allows us to estimate the probabilities of certain developmental outcomes. It does not, however, guarantee them. If the risks of a particular family environment considerably exceed the opportunities, the probability of impeded or distorted development is great. Under such circumstances, some form of intervention, to remove the risks, increase the opportunities, or both, may be required. This can also be accomplished by introducing the child to new and enriched learning environments created by professionals who are knowledgeable about the developmental needs of children. Intervention can also be accomplished by creating a healthy home environment, one that would attempt to reduce physical and health hazards and increase the type and range of stimulation and support available. The research on changing the child's family learning environment has generally gone under the rubric of "early experience" and has focused almost exclusively on the cognitive development of the child. The basic thesis is that "insufficient or improper environmental stimulation causes cognitive deficits" (Cocking, 1986). MacDonald's (1986) recent review of the literature makes it clear that the relationship is not that simple. He concludes that the data on cognitive development *do support the existence of long-range effects of early experience variables in some cases,"* but that *"intensity of ecologically appropriate stimulation in affecting behavior change"* needs to be emphasized (p. 120). Basically, cognitive processes are difficult to change and become increasingly so with age. Major and enduring changes must be initiated at a young age if there is to be an effect.

If the risks of the home environment have not been eliminated, changes are not likely to last. For these reasons, most supporters of early intervention efforts have encouraged a dual attack that both supplements the child's experiences outside the home and works towards changes in the home that can support the gains achieved externally (Bronfenbrenner, 1974).

Support for this position comes from longitudinal analyses of Head Start and similar programs (e.g., Clement *et al.*, 1984; Lazar *et al.*, 1982) and from research on the relation of home environmental variables to later intellectual functioning. For example, Elardo, Bradley, and Caldwell (1975) have studied the relationship between the child's IQ at age 3, and home environmental variables assessed at 6, 12, and 24 months of age.

Their findings show that although home environmental measures at age 6 months do not significantly relate to infant's scores on the Bayley Mental Development Scale at 6 months or 12 months of age, there is a significant relationship between these home measures at 6, 12, and 24 months and the child's Stanford-Binet IQ score at age 3 years. At age 6 months, factors related to the physical and temporal organization of the home correlated significantly with the Stanford-Binet score at age 3; at ages 12 and 24 months, the environmental factors relating to this score at age 3 were the variety of age appropriate learning materials that the mother provided and the mother's encouragement of developmental advances. These data suggest that the home-environment measures (HOME Scale) were tapping factors that were prerequisite to later intellectual development.

This research suggests that child-environment relationships may be modified, if an intensive effort to do so is initiated in a planful way and focuses on enriching the environment by increasing parental child rearing competence.

Parent-Child Relationships

Research has often linked parental expectations to children's academic success (Hess, Holloway, Dickson, & Price, 1984; Henderson, 1981). Several alternative explanations have been offered to try to explain how parental expectations and attitudes translate into child outcomes. Emphasis has been on the processes of modeling and identification with the parent, parental involvement with learning, and simple encouragement and support. Kagan and Moss (1962) proposed that children's identification with their parents causes them to imitate their parents by adopting their values and attempting to live up to their expressed expectations. Henderson (1981) combined parents' goals and expectations into a single construct called "achievement press" which he saw reflected in parental standards for school success and interest and involvement in the child's educational experiences.   The "achievement press" of parents transforms into "achievement motivation" in children. Trudewind (1982) attempted to identify specific ecological determinants of individual differences in achievement motivation, particularly in the areas of stimulation from materials found in the home, direct help with homework, stimulation from social contacts, speech training, opportunities for novel experiences outside the home, and parental achievement pressure.

Achievement press in the family system varies significantly across families. It is particularly low in those families that have a long history of operating at or below the poverty level (Belle, 1983). It is suggested that within such families the parent's feelings of powerlessness and inferiority extend to the child-school relationship and depress both achievement expectations and effort (Kamii & Radin, 1967; Hess, 1968). The thrust of many intervention programs, including Head Start has been to increase the "empowerment" of parents.

Parental expectations and a sense of efficacy in childrearing are influenced by the parents' knowledge of child development. A realistic understanding of normal development permits parents to make reasonable demands on their children and to offer appropriate stimulation.   Unrealistic demands, either too high or too low, can have deleterious effects.

Excessively high demands lead to disappointed parents and discouraged children, while excessively low expectations lead to an excessively protective environment and a lack of exploration opportunities for the child.   Both high demands and low expectations result in inappropriate levels of environmental stimulation.   The parent's ability to respond appropriately to the child's signals, a feature of the home environment that is critical for optimal cognitive development, is based on an understanding of child development (Rutter, 1985).

The research in this area suggests the need for considering both the diversity (particularly SES) of the populations studied and the nature of the parental variables (especially maternal variables) being analyzed.   The maternal variables included by Hess et al., (1984) encompassed more than the physical environment.   Two of their most powerful predictors of later school success were maternal language and childrearing practices.   Others have focused upon the important role of kinship networks and social support in determining knowledge and expectations of parents (Belle, 1983; Unger & Powell, 1980).

The focus on parental language has had a long history in the intervention literature arising from the early works of Basil Bernstein (1958; 1960; 1961) who studied the speech of low socio-econo-

mic status adults in England. Within that more strongly class stratified society, and in the days before extensive television viewing, he
documented the restrictive nature of the verbal environment of many
low-income children. This "maternal language model" of early socialization received considerable support from studies within the
United States (Hess & Shipman, 1965) and became part of an early
deficit argument for explaining why low-income children do not fare
well in the schools (Bereiter & Engelmann, 1966).

Although many of the culturally biased underpinnings of the
theoretical positions of the 1960s research have been discarded,
current empirical work continues to support the notion that parental
teaching style, as it is played out in parental language, is important.
The use of grammar and vocabulary relating to abstract concepts
and the encouragement of verbalization and feedback (Price, Hess, &
Dickson, 1981; Dickson, Hess, Miyake, & Azuma, 1979) remain important predictor variables.

Other childrearing practices have more frequently been studied
in relation to the social competence of the child. Social competence
as a construct is multi-faceted and has proven somewhat elusive in
the child development literature. However, social competence as
demonstrated by the child's independence in learning activities and
self-motivation has been shown to be positively related to cognitive
development and school success (Baumrind, 1971; Peters & Raupp,
1980).

Baumrind (1967) has identified three parental authority patterns: authoritative, authoritarian and permissive. Authoritative
parents are characterized as warm, rational, receptive to the child's
communication, and controlling, but simultaneously supportive of a
child's developing autonomy. These parents make definite behavioral
demands on the child but base them on reason rather than on authority *per se*, and encourage discussion of the issues.

In contrast, authoritarian parents are characterized as detached, controlling and less warm in their behavior. Authoritarian parents tend to demand conformity and discourage give and take. The
permissive parent is non-punitive and acceptant of the child's im-
pulses, desires and actions. Permissive parents are characterized as
non-controlling, non-demanding, and relatively warm.

In Baumrind's studies, the children of authoritative parents
fared best intellectually and had higher levels of achievement orientation and self-motivation. Authoritative parents encouraged their
children to reason, make choices, and evaluate decisions. Authoritarian parental practices inhibited participation and exploration by
the child and induced a sense of external control. The permissive
parents provided little direction or guidance for the child's behavior
which tended to decrease the child's achievement motivation.

Parents who exhibit Baumrind's warmth-hostility dimension
tended to have children characterized by achievement orientation
and academic success. This finding has been supported in research

with mothers (Hess *et al.*, 1984; Turner & Harris, 1984; Manley, 1977) and fathers (Lynn, 1974; Peters & Stewart, 1981; Lewis & Sussman, 1986).

Several variables in the parent-child relationship, therefore, seem particularly important for intervention through programs such as Head Start if the goals are to increase the child's intellectual and social competence and hence, school success. It seems important to create a discontinuity in the pattern of parent-child relationships for high risk families by bringing about changes in the areas of parental knowledge of child development, parental expectations, maternal language, childrearing practices and parental sense of efficacy. These, in turn, should produce greater encouragement for achievement and realistic levels of achievement press as well as an enriched verbal environment.

Intensity of Intervention

Little attention has been given to the intensity of the intervention within the early childhood education literature. Much of the research focusing on the impact or outcomes of participation in such programs as Head Start has not investigated or analyzed the critical features of program success.

Early research which investigated the curriculum employed did not prove particularly productive, although some of the recent follow-up studies of this research have proved intriguing (Schweinhart, Weikart, & Larner, 1986; Bereiter, 1986; Gersten, 1986). Because there was extensive confounding of program variables, only generalized conclusions about "successful" programs can be made. Successful programs have common features: careful planning and implementation, a high ratio of staff to children, a cognitively focused program with a clear framework of educational methods and goals, and at least moderate levels of parental involvement (Bronfenbrenner, 1974; Woodhead, 1985).

Second wave research in the day care field has focused attention on such variables as group size, child-staff ratios, and teacher education and experience (Belsky, 1984). Within the day care setting these variables have been found to be relatively reliable predictors of important child outcomes and they are frequently used as surrogate variables for more direct indices of program quality (Ruopp, Travers, Glantz, & Coelen, 1979).

The literature does suggest that there are important trade-offs made in designing different delivery modes for early intervention services, e.g., those made between child in-class time and hence, out of home time, and staff time spent with parents (or the parent-child dyad) to increase parental competence (Hubbell, 1983). Several studies indicate that full-day programs have a greater immediate effect than half-day programs and that full-year programs are more effective than those limited to the summer.

Furthermore, recent evidence indicates that the number of days that each child is in attendance in Head Start is associated with achievement on the Language, Math, Nature & Science, and Perceptual Scales of the Head Start Measures Battery (Bergan, 1984). On the other hand, numerous studies indicate the association between the direct involvement of the parent, usually the mother, and the short-term and long-term achievements of the children, though this line of research has not established a direct causal relationship. For example, the Head Start Synthesis Project (McKey *et al.*, 1985) reports:

* Children of parents who are highly involved in Head Start perform better on cognitive tasks than children of parents who were less involved.

* Parents involved with Head Start have more positive child-rearing approaches when interacting with their children.

Similarly, national evaluations of Home Start programs have indicated that:

* Home Start children, when compared to no-treatment controls, scored significantly higher on indices of school readiness and task orientation, but were no different from children attending more "traditional" Head Start programs.

* When mothers in the Home Start and Head Start programs were compared, the Home Start mothers spent more time in teaching readiness skills to their children and were more likely to involve their children in household tasks.

* When compared to no-treatment controls, the Home Start and the Head Start mothers taught more reading and writing skills to their children, provided more books and common playthings in their homes, read stories more often to their children, and had a higher rate of verbal interaction with their children.

Since there were no differences in child outcomes when comparing Home Start and Head Start children, it would appear that parental home teaching activities effectively compensated for the reduced in-class time the children in Home Start spent with a teacher. The literature would suggest a complementarity between in-class child time and parental in-home teaching time, with decreases in one being compensated for by increases in the other. However, there are no current data to support this linear additive model. Lacking such data, it cannot be determined whether mixed models (those involving some in-class and some parent home teaching) have the same effectiveness, less effectiveness, or greater effectiveness than more traditional center-based or home-based models. The relationship between various indices of type and intensity of intervention and subsequent child outcomes remain basically unknown.

METHODOLOGY

Head Start, Child Competence and Parental Competence

Recent research (Peters, Murphy, Bollin, & Berg, in press; Peters, Bollin, Murphy, & Berg, in press; Bollin, Peters, Murphy, & Berg, 1987) has attempted to clarify the issues of the relative efficacy of the type and intensity of intervention within Head Start programs. This research, using existing Head Start programs representing three basic delivery models: a home-based program, a traditional center-based program and mixed model using both home-visitors and center-based preschool group intervention, involved 106 Head Start children and their families. The research sought to identify variation in four basic learning opportunity variables and to relate these indicators of program intensity to both child and parent outcomes. Those variables were:

Learning Opportunities

1. <u>Child In-Class Hours</u>. This measure is the number of hours of in-center or socialization group attendance in which the child actually participated, based upon program records.

2. <u>Home Visit Hours</u>. This measure is the number of hours recorded during which a home-visitor or other paid Head Start employee was present in a particular child's home, working with the child, the parent, or both, again obtained from program records.

3. <u>Parent Volunteer Time</u>. This represents the number of hours recorded for a particular family (mothers, fathers, and other adults) devoted to specific Head Start activities. These may include participation on policy council, working in the classroom with children, helping on field trips, etc., again obtained from program records.

4. <u>Parent Home Teaching Time</u>. This represents the self-reported number of hours that parents claim to work with their children on "educational" tasks, either planned or impromptu. These data were obtained through parent interviews.

In essence, the three Head Start delivery modes were hypothesized to vary in intensity of the parent, child and joint parent-child learning opportunities they offered and such learning opportunities would, in turn, affect parent competence and child competence. The nature of these relationships is presented in Figure 1.

Home Environment Measures

Three measures of the physical characteristics of the home environment were included: toys, games and reading materials (HT), physical environment (HEN), and variety of stimulation (HVAR). Each corresponds to the subscales by the same name in the Home Observation for Measurement of the Environment (HOME) as devised by Caldwell and Bradley (1979). Data gathering and scoring procedures followed those of the original source.

Parenting Competence

Five measures of parenting competence were employed. These include:

1. <u>Child Development and Knowledge Beliefs</u>. The Knowledge of Development Scale, adapted from Duscewicz (1973), tests for knowledge of normal and atypical development. A later version of this scale (Busch, 1979) was specifically designed for parents. Items included those that tested an understand of the language and concepts of development, processes of development, and the parents' beliefs about how developmental change comes about. Scoring for the measure yielded scores for the number of items correct, the number incorrect, and the number with which the parent was unsure. Cronbach's alpha, (alpha = .68) was used to calculate reliability.

2. <u>Parental Expectations</u>. The Parental Expectations Scale, originally devised by Jensen and Kogan (1962) and modified by Busch (1979), seeks to determine parental expectations about their child's own future development. The scale is suitable for parents of handicapped and non-handicapped children and covers ten domains: self-care, education, schooling, literacy, employment and income, social interaction, mental ability, physical skills, and family management. For current purposes, only the scales for education, schooling, literacy and mental ability were used. The scales were combined and a scoring procedure was used that indicated the respondent's endorsement of HIGH aspiration items, MODERATE aspiration items and LOW aspiration items.

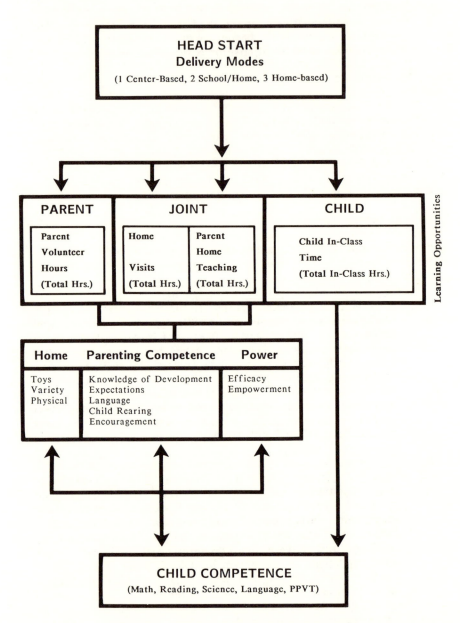

Figure 1. Model of Head Start Processes

3. Maternal Language. This measure was derived by selecting 12 items from the remaining subscales of the HOME (those not used above for assessing the home physical environment) that pertained to the mother's usage of appropriate expressive and receptive language in mother-child interactions. The scale has a Cronbach's alpha of .70.

4. Childrearing (warmth-hostility). This measure again was a derived scale of 13 items created by selection of items from the HOME scale. The Cronbach's alpha was .75.

5. Encouragement. This measure, derived again from the HOME contained 10 items and had a reliability of .75.

Power

Two "power" variables were derived from responses obtained during the open ended portion of the parent interview. In the interview, parents were asked questions, each with a follow-up, pertaining to what they felt they had gained as a result of their participation in Head Start.

1. Childrearing Efficacy. Seven of the questions pertained to the parent's knowledge of their own child's development, their role in the education of their child, their ability to deal with problems that might arise in their child's future educational experiences, and their confidence in their ability to continue to help and play a role in their child's education. Parental responses were categorized and positive categories or responses were summed across the seven questions to yield a score of the parent's perceived sense of child rearing efficacy (Cronbach's alpha = .87).

2. Empowerment. An additional eight questions focused on the parent's perceived ability to cope with family problems, their knowledge of community resources, including those for health care and nutrition, their friendship network and their sense of themselves as a person. The responses to these questions were categorized and positive responses were construed as indicating a perceived enhancement of personal efficacy, confidence, and sense of control. The total of positive responses was used as the empowerment score (Cronbach's alpha = .90).

Child Competence

Five measures of child competence were used in the current analyses. Four of these (items 2-5) were subtests of the Head Start

Measures Battery (Bergan, 1984). Detailed analyses of the measures, including information on the reliability and validity of these measures within the Head Start population, are found in Bergan, 1984.

1. General Intelligence. The Peabody Picture Vocabulary Test (Dunn & Dunn, 1981) Form L was administered to each child by a member of the assessment team. The mental age equivalent was derived using the standard scoring procedure and served in these analyses as an indicator of general intellectual ability.

2. Language. This scale taps the child's understanding of story meanings, use of words to communicate, ability to follow directions, and understanding of language rules.

3. Math. This scale includes items designed to assess the child's ability to identify and work with numbers, count, add, subtract sets, and conserve numbers.

4. Nature and Science. This scale includes items on discrimination, classification, sequencing, and prediction as well as factual knowledge about plants, the weather, etc.

5. Reading. This scale involves items on identifying and match letters and letter patterns, sentence completion, and auditory processing.

Full details of the two year longitudinal study are presented elsewhere (See Peters *et al.*, in press). For current purposes, several of the findings are highlighted.

RESULTS

Data were analyzed using a series of analyses of variance and regression techniques. All of the multivariate analyses for parents and children demonstrated significant changes over time. In accordance with the objectives of the study, however, the effects of mode--traditional, home-school, or home-based--and mode X time were the primary level of analysis. A sequential presentation of the analyses is provided to enhance the clarity of the results.

Learning Opportunities

The results of the MANOVA for the measures of learning opportunities were highly significant as were the univariate tests. The three delivery mode groups were different from one another on the class contact hours that the children engaged in. The children in the traditional, five day-week, half-day program received more supplementary learning opportunities outside the home than the other two groups, and the home-school children received more hours of supplementary learning activity than did the home-based program children. The home-school and home-based program children and parents were equivalent in the amount of home visit time they received and both groups exceeded that received by the children and parents in the traditional program. Based on self-reports of time spent, the traditional program parents and the home-school program parents spent more joint learning time in the instruction of their children than did the parents in the home-based program. Finally, the home-school program parents spent more volunteer time than did the parents in the other two programs.

Child Outcomes

Multivariate analyses of variance revealed no significant differences between the groups at any time on any of the child competence variables. Neither were any of the univariate analyses significant. That is, children in each of the three program modes made significant gains, but there were no differences among the three programs.

Parent Outcomes

Parent outcomes, as presented in Table 1, do reveal significant differences between groups at both pretest and posttest, as well as different levels of change over time. At pretest, the multivariate analysis revealed significant differences between the groups (p < .001). An analysis of the univariate results revealed that parents in the home-based program provided their children with fewer toys, games, and reading materials than other groups, but at the same time the quality of the home-based mother's language to their children was significantly better.

The groups also differed in the amount of encouragement to learn that mothers gave to their children. The traditional group demonstrated the most learning followed by the home-school group. The least amount of encouragement to learn was given by the home-based group. At posttest the multivariate analysis failed to reach significance.

Due to the exploratory nature of the research, however, and the obvious change of group levels over time (as evidenced by the failure to replicate pretest differences) univariate analyses were examined. The differences between groups in provision of toys, games and reading materials no longer existed; neither did the difference in mother's encouragement to learn. Differences were found, however, in maternal language and childrearing style, with the home-based group being rated higher than the other groups in both areas.

Table 1. Parent Variables: Pre- and Posttest Mean Scores by Mode with MANOVAs by Mode and Mode by Time

	Traditional		School/Home		Home		Pretest Multivariate $F_{(24,128)}=2.90977^c$	Posttest Multivariate $F_{(24,128)}=1.53593$	Pre- to Posttest Multivariate $F_{(24,128)}=1.89464^a$
				$N = 78$			Univariate F (x mode)	Univariate F (x mode)	Univariate F (mode x time)
	Pre	Post	Pre	Post	Pre	Post			
HOME									
Total	42.1200	44.6000	40.1852	44.4444	39.2692	48.2692			
Toys, Games, and Reading Materials	7.4800	8.3600	7.9259	8.8889	5.8846	8.8077	4.30239 (.017)^a	0.44921 (.640)	8.69380 (.000)^c
Physical Environment	5.9200	6.0800	5.5926	5.7037	5.5000	6.1538	0.32810 (.721)	0.49835 (.610)	0.71735 (.491)
Variety of Stimulation	6.3600	6.9600	6.4444	6.5926	6.2308	7.1538	0.12959 (.879)	1.20020 (.307)	1.94230 (.151)
Parent Knowledge (KDS)									
Correct	24.3600	.25.44000	25.96296	25.55556	26.30769	28.15385	1.13388 (.327)	3.01960 (.055)	2.35252 (.102)
Incorrect	6.28000	6.84000	5.96296	7.37037	5.34615	6.15385	0.66916 (.515)	1.31521 (.275)	0.60973 (.546)
Unsure	6.28000	4.72000	4.66667	4.03704	5.07692	2.65385	1.17020 (.316)	2.50167 (.089)	2.36993 (.100)

Parent's Academic Expectations

Low	0.12000	0.04000	0.11111	0.11111	0.07692	0.13980	0.19231 (.870)	1.15005 (.322)	1.27640 (.285)
Moderate	3.88000	4.00000	3.96296	3.92593	3.92308	3.92308	0.44978 (.639)	0.98228 (.379)	1.35019 (.265)
High	2.4800	2.64000	1.92593	2.18519	2.26923	2.07692	2.67949 (.075)	2.24847 (.113)	1.41003 (.251)

Maternal Competence

Maternal Language	9.64000	10.24000	9.03704	10.00000	10.42308	11.42308	3.78547 (.027)a	5.06690 (.009)b	0.43364 (.650)
Maternal Encouragement	9.04000	9.40000	9.25926	8.00000	7.34615	9.46154	5.41917 (.006)b	0.44958 (.640)	6.04571 (.004)b
Childrearing (Warmth/Hostility)	9.80000	10.16000	9.37037	10.11111	10.26923	11.50000	0.84076 (.435)	5.65404 (.005)b	0.84585 (.433)

Multivariate
F(4,142)=2.37052
(p < .055)

Parents' Feelings of Power (measurement at posttest only)

Childrearing Efficacy	7.4737		7.4783			7.6842		1.66398 (.197)	
Empowerment	6.2609		6.4783			7.3333		1.92935 (.153)	

Note: Standard deviations are in parentheses. The multivariate F is an approximate F derived from Wilks' Lambda. The multivariate analysis did not include the variable HOME total score as it was linearly dependent on other variables, producing a singular variance-covariance matrix.

a p < .05   b p < .01   c p < .001

A significant multivariate analysis by mode over time ($p < .05$) reiterated these findings, with the derived univariate results showing significant changes in the provision of toys, games and reading materials. The home-based group rose from lowest to a level equal to the others at posttest--and childrearing style--in which the home-based group appeared higher (although not significantly) at pretest and made greater gains than the other two groups.

Interrelations of Parental Competence and Child Competence

A correlational analysis of the data provided further insights into the interrelation of the variables. Significant positive relationships were found between several of the HOME subscales and measures of child intelligence and achievement. These relationships tended to be stronger across time than at either testing time. That is, HOME scores from the first testing correlated more strongly with the child measures of the second testing than with those of the first, and also showed a stronger relationship than that between parent and child measures of the second testing.

Furthermore, increases in HOME scores were positively associated with teacher hours in the home and negatively associated with the child hours in the center. Hence, greater gains in home environment variables tended to occur in cases where greater amounts of training occurred in the home. These results are entirely consistent with those of Elardo *et al.*, (1975) cited earlier.

Parallel regression analyses were employed to pursue this suggestion of different trends of influence within different program types. A separate equation was deemed necessary for each child outcome within program type (i.e., HSMB reading, language, math and science, and PPVT-mental age within each of the traditional, home-school and home-based programs) as it was hypothesized that different types of information might be acquired by the child in different manners. A separate set of five equations predicting child outcomes from the total sample represent an overall model of influences on child learning.

Predictors of Child Competence

The factors which contribute to child outcomes--as defined by the model--stem from three sources: (1) the child's ability level upon entering the program; (2) the learning opportunities provided by the program; and (3) the competencies and environmental factors associated with the parents. Since the learning opportunities in some cases were provided to parents with the objective of altering parent factors which would, based on our hypothesis, alter child outcomes, parent pretest and posttest factors were both entered into the equations. Parental feelings of childrearing efficacy and empowerment could not, however, be included in the regression due to a limited number of responses (N = 55) which would have lowered the degrees of freedom below the level deemed necessary for a meaningful analysis. The general model, then, is as follows:

Learning Opportunities + Child Pretest Scores + Parent Pretest Scores + Parent Posttest Scores = Child Outcomes

In order to derive the most meaningful equations while still retaining adequate degrees of freedom for each analysis, the following procedure was employed. First, all of the factors within each source category were regressed on each child outcome within each program and over the total sample. Secondly, those factors whose coefficients in the first level of analysis were significant at or above the .10 level were selected for entry into the final equations. All of the variables appearing in the final equations have, therefore, approached significance at the primary level of analysis and are, hence, deemed worthy of discussion regardless of their significance level in the secondary level.

The findings of these analyses indicated that the primary influence on child outcomes is child input. This was expected to be the case. Child language appeared to be the dominant factors within the traditional program, while reading, PPVT-MA and science scores appeared to explain more posttest variance in the other programs and overall. Other influential factors within the traditional program were variations in teacher contact hours in the home (despite the fact that such hours are minimal in this program) and parent pretest factors including variety of stimulation, childrearing style (p < .05), incorrect knowledge of development (p < .05), and low expectations (p < .10). A significant relationship was also found between teacher contact hours in the center and math achievement (p < .05). No parent posttest factors appeared to significantly explain child outcomes.

Within the home-school program, the non-child sources of influence were somewhat more dispersed among the source categories. Teacher contact hours in the center entered every equation except that for reading achievement and was significant (p < .05) in the explanation of language achievement. Parent pretest factors were influential primarily in the child's acquisition of mathematical knowledge, an equation which includes variety of stimulation, maternal encouragement (p < .10), correct knowledge of development (p < .05), incorrect knowledge of development (p < .10) and uncertainty about development (p < .05). The availability of toys, games, and reading materials was a contributing factor to the development of the child's understanding of science. Two parental posttest factors also entered the equations: maternal encouragement affecting science achievement and PPVT-MA, and incorrect knowledge of development affecting the acquisition of language skills.

In the home-based program, non-child influence was split

between parent pretest and posttest factors. Children's knowledge of science was affected by pretest factors including the availability of toys, games and reading materials, maternal language ($p < .01$), uncertainty about development ($p < .05$), moderate ($p < .05$) and high expectations ($p < .05$) in a time-lag fashion. Parent posttest scores, as concurrent factors which affected most child outcomes, were toys, games, reading materials, physical environment, encouragement and childrearing style. High posttest parental expectation also had a positive relationship to child development in science and PPVT-MA.

The total-sample overview of patterns of influence revealed child pretest scores to be the greatest predictor of child posttest scores. The only learning opportunity which had a relationship to a child outcome is parent volunteer hours to PPVT-MA ($p < .05$). Maternal encouragement had the most far-reaching influence, appearing in all equations, except that for math outcome. It is interesting to note that posttest maternal encouragement and toys, games and reading materials were also represented in the final full-sample equations, as these were the two areas of significant differences by mode over time according to the initial analysis by MANOVA with the home-based program showing the greatest increases.

Predictors of Parental Competence

The Head Start program is based on the assumptions that (1) parent beliefs and behaviors affect child development, and (2) positive changes in parent behaviors may be induced by learning opportunities, and (3) that these changes can, in turn, enhance child development. The analysis of predictors of child outcomes indicated that parenting factors were the second most influential source of prediction in the model. However, it was deemed necessary to examine the determinants of parent posttest scores to better validate the given assumptions. The general model is as follows:

*Learning Opportunities + Child Pretest Scores +
Parent Pretest Scores = Parent Posttest Scores*

The final regression equations predicting parent outcomes were derived in a manner similar to that employed to predict child outcomes: first the sources of influence (learning opportunities and parent and child input) were regressed on parent posttest scores separately, and then those factors contributing significantly to the first level equations were entered into the final equations. Over the total sample, maternal encouragement was found to have a positive relationship with teacher contact hours in the home and with reported hours of home instruction ($p < .10$).   Childrearing style (warmth-hostility) was found to have a relationship with teacher contact hours in the center as did toys, games and reading materials with reported hours of parent instruction, probably because the most positive change in those areas was shown by the parents of the home-based program who received the least in-class time and reported the lowest hours of instruction.

Within the home-based program, hours of teaching in the home predicted the posttest quality of the physical environment (p < .10) and was the only significant predictor of mothers' posttest language (p < .05). Parent volunteer hours contributed to the prediction of maternal instruction while reported hours of parent instruction exhibited a positive relationship to both the provision of toys, games and materials as well as high expectations of parents for their children.

Within the traditional program, teacher hours in the center was related to maternal language and encouragement to learn. Parent hours of instruction, however, was positively related to posttest levels of physical environment, variety of stimulation, maternal language (p < .05), childrearing style (p < .10) and moderate low expectations. These areas at pretest were related to child posttest scores, but not at posttest.

Child-related behaviors of parents within the home-school program did not seem to be influenced by Head Start programming as the only relationship found was between reported hours of instruction and maternal encouragement to learn, indicating a sort of inter-rater reliability on judgments of the same behaviors by both parents (reported hours) and home visitors (maternal encouragement).

DISCUSSION

The most obvious conclusion from the data presented is that parents are more likely to show differential short-term effects from intervention than are children. This is possibly a function of children's scores measuring the acquisition of new concepts and skills while parents' scores, in most cases, measure a restructuring of existing concepts and skills, changing their emphasis or perspective.

Children within each of the three programs made gains in all of the measured outcomes despite the between-group variability of in-class instruction. Therefore, either these gains were unrelated to classroom instruction, or as hours of in-class instruction decline, other factors--presumably in-home instruction and reported hours of parent instruction--compensate for their effect. The trade-off, however, cannot be an equivalent one in terms of average overall hours of child instruction (in-class hours + home visit hours + parent instruction hours) as these total hours differ greatly between modes: traditional - 819 hours; home-school - 619 hours; school - 221 hours. It would appear by this simple logic, that the type of instruction provided by the home-based program is far more effective in producing child gains.

This explanation, however, assumes that the short-term child gains reported are a result of instruction, which is questionable in view of the regression analyses discussed. Rather, the regression analyses reveal a very complex interplay of type and amount of instruction, parent characteristics and child characteristics which vary predictably by type and intensity of intervention.

In the traditional program, where the intervention focuses primarily on the child in a school context, child gains appear to derive from that formal schooling. The amount of time parents report spending in instruction of their children declines from pretest to posttest, possibly indicating an abrogation of responsibility for the role of "teacher." Within the traditional program, the learning opportunities provided by the Head Start program (teacher hours in context and teacher hours in home) demonstrate no positive effects on maternal behavior and, in fact, were found to be associated with a decrease in the levels of maternal language and encouragement to learn. No effect of mothers' behaviors at posttest were found, once the effects of behavior at pretest were removed. While the direct effects of the program on the children may have been positive, the indirect effects, which could logically ensue from intervention focused on the parents, was negligible or negative.

Within the home-school program, where child and parent are separate targets of intervention, the program-imposed learning opportunities affected child gains and, to a lesser degree, parental change. The parents' hours of reported instruction increased. The increase was further reflected in an increase in the posttest measuring of maternal encouragement to learn which, in turn, had a greater effect on childrens' achievement at posttest than at pretest levels. Hence, the benefits to the child may be two-fold in this program, reflecting both concurrent and time-lagged influences.

The home-based program, which concentrates on improving the academically-related interaction of parent and child, appeared to demonstrate its effectiveness in a different manner. Although the formal Head Start intervention did not appear to influence child achievement directly, it did seem to significantly affect the way parents structure the child's environment and the availability of learning materials, as well as the level of mothers' language and encouragement to learn. These factors, in turn, did affect the rate of achievement progress shown by the children.

Over the total sample, the pattern of effect most closely mirrors that of the home program with direct child instruction having negligible influence and increases in mothers' encouragement to learn contributing consistently to child achievement in all areas but math.

The number of hours of each type of teacher-contact was a factor predicting within program effectiveness. Variance in child in-class hours did not affect the home program where it was minimal, and affected only math learning in the traditional program, with its high average number of hours. Variance was a factor within the home-school program, however, possibly indicating a threshold effect wherein a two-day program provides the minimum amount of time necessary to show effects, while the traditional program is well over the required threshold and the home-based program well under. The same pattern appeared in reversed direction in the effects of home visitor hours,with this variable demonstrating an effect in the program where it was lower, but not in the programs where its average level is higher. Thus, both the quantity and focus of intervention are reflected in the patterns by which young children acquire knowledge.

On this relatively short-term comparative intervention, all of the children appear to have made comparable gains, while differential gains were made by the parents. Therefore, the only conclusion which can logically be drawn from the data at this point in time is that variations in mode of Head Start delivery do not significantly affect the rate of short-term gains in achievement in children. But the question is raised as to what the long-term effects of the differential treatments will be.

In the traditional program a discontinuity has been established between the child's school and home environments. As the focus of training shifts, however, to the parent-child dyad, continuity of environmental stimulation, expectations, and encouragement is established at a new level higher than the child's original environment. As parents work with their children and become better able to assess their development and establish reasonable expectations, they appear to be reinforced by recognition of the child's progress and realization of those expectations. Parents involved directly with their children tended to show higher levels of feelings of child-rearing efficacy and general empowerment in dealings with their families and community services. Their children, in turn, seemed to benefit from continuous reinforcement and consistent, realizable expectations.

It may be, in light of these patterns of interaction, that the durability of effects of intervention is determined by the incorporation of the procedures and values espoused by the intervention into the system of natural consequences which compose the child's enduring environment. While it was assumed by early intervention programs that introducing discontinuity into the "deprived" child's environment would be beneficial, redefined continuity may be a more potent key to enhancing children's academic performances. ▓

This research and the preparation of this chapter were funded by a grant from the Head Start Bureau, Administration on Children, Youth and the Family, U.S. Department of Health and Human Services. The author wishes to express his appreciation to Allen N. Smith, ACYF, Head Start Bureau, for comments on earlier draft materials and to the many Head Start staff and participants who contributed time to this study.

REFERENCES

Baumrind, D. Current patterns of parental authority. *Developmental Psychology Monographs*, 1971, 4(1, Pt. 2).

Baumrind, D. Child care practices anteceding three patterns of preschool behavior. *Genetic Psychology Monographs*, 1967, 75, 43-88.

Belle, D.E. The impact of poverty on social networks and supports. In L. Lein & M.B. Sussman (Eds.), *The ties that bind: Men's and women's social networks.* New York: Haworth Press, 1983.

Belsky, J. Two waves of day care research: Developmental effects and conditions of quality. In R.C. Ainslie (Ed.), *The child and the day care setting: Qualitative variations and development.* New York: Praeger, 1984.

Bereiter, C. Does direct instruction cause delinquency? *Early Childhood Research Quarterly*, 1986, 1, 289-292.

Bereiter, C., & Engelmann, S. *Teaching disadvantaged children in the preschool.* Englewood Cliffs, NJ: Prentice-Hall, 1966.

Bergan, J.R. *The Head Start measures project: Path-reference assessment for Head Start children, executive summary.* Tucson: The University of Arizona, 1984.

Bernstein, B. Some sociological determinants of perception. *British Journal of Sociology*, 1984, 9, 159-174.

Bernstein, B. Language and social class. *British Journal of Sociology*, 1960, 11, 271-276.

Bernstein, B. Social structure, language and learning. *Educational Research*, 1961, 3, 163-176.

Bollin, G., Peters, D., Murphy, R., & Berg, M. Parental child rearing competence and children's school competence in low-income families. Paper presented at the annual meeting of the American Educational Research Association, Washington, D.C., 1987.

Bradley, R., & Caldwell, B.M. The relation of infant home environments to mental test performance at fifty-four months: A follow-up study. *Child Development*, 1976, 47, 1172-1174.

Bronfenbrenner, U. *A report on longitudinal evaluations of preschool programs. Vol. 2: Is early intervention effective?* Office of Child Development. DHEW Publication No. ([OHD]74-24). Washington, D.C.: Government Printing Office, 1974.

Busch, N.A. Parental development in mothers of handicapped, at-risk and normal children. Unpublished doctoral dissertation. The Pennsylvania State University, 1978.

Caldwell, B., & Bradley, R.H. *Home observation for measurement of the environment.* Little Rock: University of Arkansas, 1979.

Clement, J.R., Schweinhart, L.J., Barnett, W.S., Epstein, A.S., & Weikart, D.P. *Changed lives: The effects of the Perry Preschool Program on youths through age 19.* Ypsilanti, MI: High-Scope Press, 1984.

Cocking, R. The environment in early experience research: Directions for applied developmental investigations. *Journal of Applied Developmental Psychology,* 1984, 7, 95-99.

Dickson, W.P., Hess, R.D., Miyake, N., & Azuma, H. Referential communication accuracy between mother and child as a predictor of cognitive development in the United States and Japan. *Child Development,* 1979, 50, 53-59.

Dunn, L.M., & Dunn, J.A. *Peabody picture vocabulary test: Manual.* Circle Press, MN: American Child Guidance Services, 1981.

Duscewicz, R. The knowledge of development scale. *Education,* 1973, 93, 252-253.

Elardo, R., Bradley, R.H., & Caldwell, B.M. The relation of infants' home environments to mental test performance from six to thirty-six months: A longitudinal analysis. *Child Development,* 1975, 46, 71-76.

Garbarino, J. *Children and families in the social environment.* New York: Aldine, 1982.

Gersten, R. Response to "Consequences of three preschool curriculum models through age 15." *Early Childhood Research Quarterly,* 1986, 1, 293-302.

Henderson, R.W. Home environment and intellectual performance. In R.W. Henderson (Ed.), *Parent-child interaction: Theory, research and prospects.* New York: Academic Press, 1981.

Hess, R.D. Early education as socialization. In R.D. Hess & R.M. Bean (Eds.), *Early education: Current theory, research and action.* Chicago: Aldine, 1968.

Hess, R., Holloway, S., Dickson, W.P., & Price, G.G. Maternal variables as predictors of children's school readiness and later achievement in vocabulary and mathematics in sixth grade. *Child Development,* 1984, 55, 1902-1912.

Hess, R.D., & Shipman, V.C. Early experience and the socialization of cognitive modes in children. *Child Development,* 1965, 36, 869-886.

Hubbell, R. *A review of Head Start research since 1970.* Washington, D.C.: CSR, Inc., 1983.

Jensen, G.D., & Kogan, K.L. Parental estimates of the future achievement of children with cerebral palsy. *Journal of Mental Deficiency Research,* 1962, 6, 56-64.

Kagan, J., & Moss, H.A. *Birth to maturity, a study of psychological development.* New York: Wiley, 1962.

Kamii, C.K., & Radin, N.L. Class differences in the socialization practices of Negro mothers. *Journal of Marriage and the Family,* 1967, 29, 302-310.

Lazar, I., Darlington, R., Murray, H., Royce, J., & Snipper, A. *Lasting effects of early education.* Monographs of the Society for Research in Child Development, 1982, 47, (1-2, Serial No. 194).

Lewis, R.A., & Sussman, M.B. *Men's changing roles in the family.* New York: Haworth Press, 1986.

Lynn, D.B. *The father: His role in child development.* Monterey, CA: Brooks/Cole, 1974.

MacDonald, K. Early experience, relative plasticity, and cognitive development. *Journal of Applied Developmental Psychology*, 1986, 7, 101-124.

Manley, R.O. Parental warmth and hostility as related to sex differences in children's achievement orientation. *Psychology of Women Quarterly*, 1977, 3, 229-246.

McCubbin, H., Sussman, M.B., & Patterson, J. (Eds.), *Social stress and the family.* New York: The Haworth Press, 1983.

McKey, R.H., Condelli, L., Ganson, H., Barrett, B.J., McConkey, C., & Plantz, N.C. The impact of Head Start on children, families and communities, executive summary. Washington, D.C.: CSR, Inc., 1985.

Peters, D., Bollin, G., Murphy, R., & Berg, M. An analysis of three modes of Head Start delivery. *Early Childhood Research Quarterly* (in press).

Peters, D., & Kontos, S. Continuity and discontinuity of experience: An intervention perspective. In D.L. Peters & S. Kontos (Eds.), *Continuity and discontinuity of experience in child care.* Norwood, NJ: Ablex Publishers, 1987.

Peters, D., Murphy, R., Bollin, G., & Berg, M. Head Start as a means of creating a family environment conducive to literacy. In S. Silvern (Ed.), *Literacy through family, community and school interaction.* JAI Press, (in press).

Peters, D.L., & Raupp, C.D. Enhancing the self-concept of the exceptional child. In T. Yawkey (Ed.), *The self-concept of the young child.* Provo: Bingham Young University Press, 1980.

Peters, D.L., & Stewart, R.B. Father-child interactions in a shopping mall: A naturalistic study of father role behavior. *The Journal of Genetic Psychology*, 1981, 138(2), 269-278.

Price, G.G., Hess, R.D., & Dickson, W.P. Processes by which verbal-educational abilities are affected when mothers encourage pre-school children to verbalize. *Developmental Psychology*, 1981, 17, 554-564.

Ruopp, R.R., Travers, J., Glantz, F., & Coelen, C. *Children at the center: Final report of the national day care study.* Cambridge, MA: Abt Books, 1979.

Rutter, M. Family and school influences on cognitive development. *Journal of Child Psychology and Psychiatry*, 1985, 26, 683-704.

Schweinhart, L.J., Weikart, D.P., & Larner, M.B. Consequences of three preschool curriculum models through age 15. *Early Childhood Research Quarterly*, 1986, 1, 15-46.

Turner, P.H., & Harris, M.B. Parental attitudes and preschool children's social competence. *The Journal of Genetic Psychology*, 1984, 144, 105-113.

Trudewind, C. The development of achievement motivation and individual difference: Ecological determinants. In W.W. Hartup (Ed.), *Review of child development research.* Chicago: University of Chicago Press, 1982.

Unger, D., & Powell, D. Supporting families under stress: The role of social networks. *Family Relations*, 1980, 29, 566-574.

Woodhead, D. Pre-school education has long-term effects--but can they be generalized? *Oxford Review of Education*, 1985, 11, 133-155.

▨ 8 ▨

FROM LATCHKEY STEREOTYPES TOWARD
SELF-CARE REALITIES

Hyman Rodman

INTRODUCTION

A tremendous surge of interest in school-age child care has arisen within the past few years. This includes state and federal efforts for legislation to support school-age child care, as well as community efforts to set up after-school programs and to organize telephone help services for children who spend time alone. Many of these efforts are based on the presumed negative consequences of the so-called latchkey arrangement. However, very little is known about the consequences of the latchkey arrangement for children's functioning.

Does it contribute substantially or marginally or not at all to developmental problems such as anxiety and fear, or to developmental advances such as responsibility and maturity? What aspects of the arrangement or the context in which it occurs contribute positively or negatively to a child's development? Research data are not available to answer these questions. The limited evidence raises serious questions about whether the latchkey arrangement has negative consequences (Rodman, Pratto, & Nelson, 1985) and about the process by which it has come to be seen in negative terms.

It is chastening to recall the example of the 1950s and 1960s when many people believed that day care and working mothers jeopardized children's development. Subsequent research indicated that these beliefs were wrong and that many intervening variables were important (D'Amico, Haurin, & Mott, 1983; Etaugh, 1980; Hoffman, 1979). Unfortunately, professional and lay concern about latchkey children in the 1980s seems to be repeating the earlier history on day care and working mothers.

99

Clearly one of the most difficult problems faced by working parents is finding adequate and affordable child care, and they often have to juggle schedules and make frantic arrangements in order to manage. More child care programs would most certainly help this situation. The argument for such programs, however, should be based on demographic and labor force changes and on parents' and children's needs for reliable services (Zigler & Gordon, 1982; Rodman & Cole, 1987). They should not be based on unsubstantiated claims of negative consequences.

The conflict between the scientist's desire to know more about a topic and the practitioner's need to deal with practical problems is pandemic. Such conflict stems from differences in professional training and orientation, and can be found in the physical, biological, and social science arenas. The conflict is sometimes exacerbated by the media, which are often ready to rely on the statements of one scientist or the findings of an unpublished study. Recent media reports about latchkey children illustrate the problem. Although there is a fairly strong scientific norm that findings should initially be released through peer-reviewed journals, that norm is not shared by all researchers, nor is it shared by the press. As a result, many press reports are based on data or opinions that have not been reviewed or validated by the scientific community. Furthermore, the members of the press frequently find it difficult to evaluate the credibility of the scientific information they receive (Atkins & Rivers, 1987).

THE POSITIVE AND NEGATIVE CONSEQUENCES OF SELF-CARE

Some educators and social scientists have referred to the positive consequences of self-care, but the preponderance of commentary has referred to risks, dangers, and negative consequences. The following brief review points out how little is known.

Long and Long (1982; 1983) have carried out research comparing self-care, sibling care, and adult care children from grades one through six. They report higher levels of fear, loneliness, and boredom among self-care children. But the Longs have not attempted to control any variables that might influence their findings. Nor have they asked the same questions of children in these different groups. As a result, it is difficult to know how much confidence to place in their research. They themselves have candidly acknowledged the possibility of bias due to the subjective methods they have employed (Long & Long, 1982).

In a rural-area study of fifth- and seventh-graders, Galambos and Garbarino (1983) compared 21 self-care children, 27 children cared for by a mother, and 29 adult-care (other than mother) children. They found no significant differences on measures of school adjustment, classroom orientation, achievement level, or fear level. Woods (1972) compared adult-supervised and unsupervised fifth-grade children in Philadelphia and found few significant differences, but unsupervised girls did have significantly lower scores on measures of school achievement and school relations.

In a study of matched samples of children in self-care and adult care arrangements in a North Carolina community, there were no significant differences on measures of self-esteem, locus of control, or teachers' reports on personal and social adjustment. The sample consisted of 26 matched pairs of children in the fourth grade and 22 matched pairs in the seventh grade--a total of 48 matched pairs of 96 children (Rodman, Pratto, & Nelson, 1985).

Three or four studies, of course, cannot answer the question of whether there are negative or positive consequences of self-care. Many variables need to be controlled: socioeconomic status, race, sex of the child, nature of the parent-child relationship, amount of time the child spends in self-care, and whether the family has selected self-care voluntarily or involuntarily. Despite the limited evidence, numerous reports are being circulated about the disadvantages of self-care. Congressional testimony includes statements implying that the negative consequences of self-care are well-established facts (Flynn & Rodman, In press).

Latchkey Stereotypes

Although the terms "latchkey children" and "unsupervised children" are frequently used, it is best to avoid them because of the negative connotations associated with the terms. Although it is an option used by many families while the parents are working, some argue that leaving a child at home alone is not a child care arrangement at all, but the absence of an arrangement. And the terms "latchkey" and "unsupervised" can conjure up images of predelinquent or neglected children. In the second place, these terms are inaccurate. Some self-care children care for themselves before school; some care for themselves in the evening hours. In these cases they do not need a key and the "latchkey" term is inappropriate. Moreover, many self-care children are well-supervised, albeit from a distance. Parents of many of these children have worked out a variety of supervisory rules and regularly talk to their self-care children by telephone--so that the term "unsupervised" does not seem appropriate either (Rodman, Pratto, & Nelson, 1985).

Although there is much to learn about the realities of self-care, some of the stereotypes about latchkey children are quite clear. Several possible factors contribute to the strong negative picture painted of latchkey children.

First, for anyone doing research on or developing programs for latchkey children, there is an incentive to see a problem. An urgent problem is likelier to loosen the purse strings of government agencies or private foundations, either to learn more about the nature of the problem or to do something about alleviating it.

Second, the media have a vested interest in problems. As Norman Ornstein has said, the media are *"biased in favor of negative news and...conflict and controversy"* (cited in Hunt, 1985). The media also love dramatic photographs and drawings: children with keys around their necks or children cowering in closets. These add appeal to human interest stories. They sell newspapers and books. A few examples will illustrate the stereotypes of latchkey children presented by the media:

▓ The *New York Times*, April 4, 1984. A story by reporter Robert Lindsey refers to the high percentage of families in which both parents work, and immediately follows this with a quotation from a detective in the Los Angeles Police Department: *"You're looking at the latch-key children, the kids who are left alone or with someone who abuses them."* The story continues with accounts of sexual abuse of nursery school children.

▓ The *Christian Science Monitor*, April 28, 1981. Editor Curtis J. Sitomer starts the first of a series of four articles on "Who Cares for the Children?" this way: *"For eight-year-old Angie, days are often long, lonely and sometimes scary."* Later he continues: *"Angie is a 'latchkey' child--a term defining a child who has a house key dangling from a chain around the neck and trudges home after school to an empty house. Sometimes a parent has set up an agenda for homework, house chores, and leisure time activities. More often there is none. Then television becomes the No. 1 child minder. The lure of the streets and delinquent or predelinquent activities too frequently fill the vacuum....At worst the latchkey child is a heartrending case of abuse and abandonment--a ticking time bomb for the future. Even at best, the latchkey child soberingly reminds an adult of those marginal degrees of loneliness, idleness, and jeopardy that can afflict even the most affluent home when no parent is there."*

▓ The *New York Times*, November 15, 1983. Fred M. Hechinger, in a reasoned statement about the need for more school services for children, says: *"With more than 22,000 daycare centers already in operation across the country, the trend is obvious and ques-*

*tions about need seem irrelevant. So, in the face of
armies of latchkey kids, is the question whether
children need safe and congenial places to spend their
after-school hours. Those who denounce organized
day care and after-school programs as an usurpation
of the home's obligations are out of touch with the
realities of today's home."*

In short, the media give us the lonely latchkey child trudging
home alone, and armies of latchkey children roaming the streets.
They equate latchkey children with delinquent and pre-delinquent
activities, with idleness and fear, with abuse and abandonment.

There are also balanced stories in the media, but the reports
are heavily weighted toward negative portrayals of the latchkey
situation.

CONCLUSION

Latchkey stereotypes are common, but very little is actually
known about the reality of the self-care situation and its consequences for children. Little is known about what children do
while they are in self-care, about their attitudes and feelings toward
self-care, about the rules and regulations that parents establish for
them, and about the amount of time they spend in self-care.

Much more also needs to be learned about the consequences of
the self-care arrangement for children. The question is complicated
by ad hoc changes parents make in their child care arrangements to
deal with emergencies, and by changes over time as children grow
older and as the family and work situation change. The presence or
absence of siblings as child caretakers, the age and competence of
caretakers, the age and competence of the self-care children, and
the relationship between caretaker and child are also complicating
factors. The social and legal context of the arrangements is also
important (Sussman, 1983).

In short, as with the consequences of working mothers or day
care, learning about the consequences of self-care is a complex issue
involving many interacting variables (Murray, 1983; Sigel, 1983). It
is risky to come to premature conclusions about the consequences of
self-care, and it is important to be alert to the possibility that some
of the early negative reports are not sufficiently cautious about the
complexity of the situation. ▓

This is a revised version of a paper presented at the First National
Conference on Latchkey Children, Boston, May 17, 1984. Thanks are
due to the William T. Grant Foundation and the Ford Foundation for
their support of our research on self-care children.

REFERENCES

Atkins, G., & Rivers, W. *Reporting with understanding.* Ames: Iowa State University Press, 1987.

D'Amico, R.J., Haurin, R.J., & Mott, F.L. The effects of mothers' employment on adolescent and early adult outcomes of young men and women. In C.D. Hayes & S.B. Kamerman, (Eds.), *Children of working parents: Experiences and outcomes.* Washington, D.C.: National Academy Press, 1983.

Etaugh, C. Effects of nonmaternal care on children: Research evidence and popular views. *American Psychologist,* 1980, 35, 309-319.

Flynn, C., & Rodman, H. Latchkey children and afterschool care: A feminist dilemma? *Policy Studies Journal,* (In press).

Galambos, N.L., & Garbarino, J. Identifying the missing links in the study of latchkkey children. *Children Today,* 1983, (July-August) 2-4, 40-41.

Hoffman, L. W. Maternal employment. *American Psychologist,* 1979, 34, 859-865.

Hunt, A.R. Media bias is in eye of the beholder. *Wall Street Journal,* July 23, 1985, 32.

Long, T.J., & Long, L. *Latchkey children: The child's view of self care.* Washington, D.C.: Catholic University of American (ERIC Document Reproduction Service No. ED 211 229), 1982.

Long, T.J., & Long, L. *The handbook for latchkey children and their parents.* New York: Arbor House, 1983.

Murray, T.H. Partial knowledge. In D. Callahan, & B. Jennings, (Eds.), *Ethics, the social sciences, and policy analysis.* New York: Plenum Press, 1983.

Rodman, H., & Cole, C. Latchkey children: A review of policy and resources. *Family Relations,* 1987, 36, 101-105.

Rodman, H., Pratto, D., & Nelson, R. Child care arrangements and children's functioning: A comparison of self-care and adult-care children. *Developmental Psychology,* 1985, 21, 413-418.

Sigel, I.E. The ethics of intervention. In I.E., Sigel, & L.M.Loasa (Eds.), *Changing families.* New York: Plenum Press, 1983.

Sussman, M.B. Law and legal systems: The family connection. *Journal of Marriage and the Family,* 1983, 45, 9-21.

Woods, M.B. The unsupervised child of the working mother. *Developmental Psychology,* 1972, 6, 14-25.

Zigler, E.F., & Gordon, E.W. (Eds.), *Day care: Scientific and social policy issues.* Boston: Auburn House, 1982.

▓ 9 ▓

MISERS AND WASTRELS: PERCEPTIONS OF THE DEPRESSION
AND YUPPIE GENERATIONS

Irwin Deutscher

INTRODUCTION

There is a tradition among many social scientists of identifying
with the underdogs and rallying our knowledge to their support
(Becker, 1961). I concur with the spirit of such efforts, but, as
Alvin Gouldner (1962) observed, it is not always a simple matter to
recognize the underdog and, therefore, to know whose side
compassion compels us to join. This paper will consider the
economic condition, world view and perceptions of others held by
the elderly as well as the perceptions some younger people hold of
the elderly. Are older people the categorical underdogs they are
sometimes thought to be? There is no intent here to deny respect
for those among the elderly who have earned it or to deny the
social and economic responsibility a society has for those among the
elderly who are in need.

Three kinds of evidence were considered in this chapter.
Somewhat reluctantly, for reasons which will be discussed later,
certain numbers and rates which were purported to be objective
"facts" provide some of the evidence employed here. A second
source of evidence was an analysis of the historical milieu in which
these generations matured. The final kind of evidence was the most
persuasive. It consisted of mutual perceptions held by two
generations of the American upper-middle classes--largely as
reflected in the mass media. Under consideration here are the
consequences of according economic benefits and other privileges to
a segment of society *on the basis of age alone*. Further
consideration is given to which segments of the society see
themselves as providing those benefits and privileges. For the pur-

poses of this paper, we will attempt to set aside for the moment, the complexity surrounding the Social Security System in the United States and, instead, to emphasize certain other practices that are less complicated.[1]

Statistics of Age and Income

What do the numbers suggest about the relationship between age and poverty? The most recent analysis available at this writing is a report from the Social Security Administration released in September, 1986 (Radner, 1986).   Radner's data show a dramatic decrease in poverty rates for older persons between 1967 and 1983. Furthermore, after age 55, the older the person, the greater the decline in poverty rate with the extreme age group (age 85 and over) having its poverty rate reduced to half of what it was in 1967 (Radner, 1986:14, Fig. 6).   This dramatic reduction of poverty rates through those years persists for elderly "persons," "family units," "families," and "unrelated individuals." During the same period there is an increase in the poverty rate for younger "persons," "family units," and "families" (Radner, 1986:44, Table 31).

Radner also provides a breakdown comparing the relatively poor (2nd income decile) with the relatively rich (9th income decile) (Table 6, p. 22).[2] We find here no change between 1967 and 1985 in the overall income of the relatively poor who were under age 65. The mean income of those relatively poor over 65, however, increased to the point where by 1983 it was nearly the same as the average income for those under 65 ($4,800 vs. $4,900 per year in constant 1983 dollars).   Although the more prosperous group of elderly showed higher percentage increase in income over those years than did their younger counterparts, those relatively rich persons under age 65 maintained a considerable lead in dollar income in 1983 ($23,030 vs. $28,210).

An effort to encapsulate some of Radner's findings appears in Table 1.   During the earlier years, it was among the younger families that income increased at nearly three times the rate of that of the older ones. But the reverse occurs in the later years where those family units headed by persons over 65 increased their income by more than three times the rate of the younger family units.

Gordon Streib summarizes a different type of data. He informs us that in the late seventies over 80% of elderly couples in the United States lived in a home they owned and over 80% of these owned them free of debt (Schultz, 1980 cited by Streib, 1985: 345). This suggests something less than overriding poverty among elderly people especially when one considers the large amounts of tax free equity in those homes.   Observe, however, that less than half this percentage of "elderly single persons" live in an owned home. "Elderly single person" is, I suspect, by and large a euphemism for old women.   Although most old women may not be poor, there is

Table 1.   Percentage Change in Real Mean Income of Family
          Units, By Age of Unit Head, 1947-83[a]

| Time Period | Percent Changed (all +) | | |
	Under Age 65	Over Age 65	All Ages
1947–1967	68	26	59
1967–1983	12	40	15

[a]Income has been converted to constant 1983 dollars.
Source: A fragment of Table 1, pg. 18 of Radner (1986)

evidence that the probability of being poor is considerably greater among older women than among old people in general.

If we are to allocate resources among categories of the population in which need is concentrated, then older women would seem a more reasonable category than old people generally. Various representatives of the Older Women's League made pronouncements to this effect in their report on Mother's Day 1986. The statistics presented there are summarizes in the statement by Representative Mary Rose Oakar of Ohio: "To be seventy years old and female is to be alone and poor" (The Washington Post, 1986). Susan Grad (1984) reports dollar income data from 1960 to 1982 by age and family configuration (See Table 2, Grad, 1984 for a summary of these findings). Among the elderly it appears that the highest percentage of persons living in poverty is indeed women living alone. Note however that there is one category among the "nonaged" which has an even higher proportion living in poverty.

In Table 2 we see that the category in which the highest percentage of incomes below the poverty line in 1982 is among families headed by "nonaged" women. Furthermore, it has been a decade (1972) since elderly women exceeded the poverty rate of younger women. It is no surprise that the money income of families headed by women is less than that of those headed by men. But it is also true that women in age groups under 45 have all seen a decrease in their dollar incomes between 1970 and 1982. In contrast, families headed by women 65 and older not only had higher incomes than those under 45 in 1970 but have increased rather than decreased their incomes in 1982 (Grad, 1984:15, Table 14). These data suggest that there is a category of younger women with dependent children which is more needy than older people in general or even than older women in general. Could this be the underdog category most deserving advocacy?

Table 2. Percent of the Aged and Nonaged with Incomes Below the Poverty Line for Various Economic Units at 2-year Intervals, 1960-82

Age, economic unit and sex	1960	1962	1964	1966	1968	1970	1972	1974	1976	1978	1980	1982
Aged												
Families and unrelated indiv.:												
Families	27	26	23	21	17	16	12	8	9	8	9	9
Headed by men	26	25	22	21	16	16	11	8	8	8	8	8
Headed by women	31	31	17	20	22	20	16	12	14	12	15	16
Unrelated individuals	66	62	61	54	49	47	37	32	30	27	31	27
Men	60	53	50	44	44	39	26	27	26	21	24	21
Women	68	65	65	57	51	50	40	33	32	29	32	29
Persons, total[a]	35	(1)	(1)	(1)	(1)	24	19	16	15	14	16	15
Men	(1)	(1)	(1)	(1)	(1)	19	13	12	11	10	11	10
Women	(1)	(1)	(1)	(1)	(1)	28	22	18	18	17	19	18
Nonaged												
Families and unrelated indiv.:												
Families	16	15	13	10	8	8	8	8	9	8	10	12
Headed by men	13	12	10	7	6	6	5	5	5	5	6	7
Headed by women	44	44	27	35	33	32	32	33	32	31	32	31
Unrelated individuals	32	33	30	26	22	20	20	19	19	17	17	19
Men	26	28	24	21	16	14	15	16	15	13	14	15
Women	38	38	35	30	27	25	25	22	23	21	21	22
Persons, total[b]	(1)	(1)	(1)	(1)	(1)	9	8	8	8	8	9	12
Men	(1)	(1)	(1)	(1)	(1)	7	6	6	6	6	7	9
Women	(1)	(1)	(1)	(1)	(1)	10	10	10	10	10	11	14

[a] Not available.
[b] Nonaged persons are those aged 22-64.
Source: Table 7, pg. 10 in Grad (1984).

Finally, there is another category which is sometimes put forth as the poorest of them all and that is people under age 16 or 17. A Wall Street Journal article attributing its source to the U.S. Census Bureau presents data of age and poverty suggesting that between 1970 and 1985 people age 17 and under represent an increasing proportion of the poor while those age 65 and over are decreasingly poor (1/13/87, p. 35). In March of 1986, Ellen Goodman reported in her column that "Today a child in America is six times more likely to be poor than an elderly person" (The Washington Post). This comment is repeatedly made, frequently citing Preston's 1986 Presidential Address to the Population Association of America.

At least for now, this is an unclear statistic, since most children have no wealth of their own and no child controls any wealth. Nevertheless, it appears that as an age category, children are more likely than others to be found in families below the poverty line. If we are searching for a statistically needy category, old women, women with dependent children, and possibly children in general are more likely to be poor than are old people generally. Yet, consider that when we treat these age/sex segments categorically, we are including within each of them a majority which is not living below the official poverty line.

In Table 2, the highest percentage of people below the poverty line is found among women under age 65 who are heads of households. But even in this most economically deprived category, 69 percent of the women have incomes above the poverty line. An age category remains an imprecise and not always just criterion for allocating resources. Such rates may help us to locate specific segments of the population where need is concentrated but they also incline us to provide assistance for the majority in the category who may not need it. Furthermore, such categories lead us to neglect the many poor people in categories with low poverty rates. It is necessary to remember that such rates may have nothing to do with numbers. It is possible that there may be more poor people in categories with low poverty rates than in categories with high ones. This suggests that something other than age per se of age/sex categories is needed to provide a just reallocation of society's resources.

All of the statistics cited above must be considered with caution. Authors such as Radner and Grad are convinced that their data is ambiguous and can be read in various ways depending upon the perspective of the analyst. It is not clear who are the poorest in America. There are, however, certain demographic statistics which are less ambiguous. There was in fact a "baby boom" following World War II and that little blip in the population pyramid continues to age. Among these so called "boomers" is an upper middle class group which we refer to (as does the press) as YUPPIES. But it is also true that the percent of all family units headed by persons over and under age 65 is nearly identical in 1983 to what it was in 1967 (Radner, 1986:20, Table 4). One can project these figures with some accuracy into the 21st century and it is unlikely that such projections will suggest any dramatic age redistribution of the population.

Other reliable statistics suggest that Americans are choosing to retire earlier as the years go by. For example, in 1974 the average age of Ohio teachers retiring was 63. Ten years later the average retirement age had dropped to 61. As of June, 1985 the average retirement age was 59. For the national TIAA-CREF (a private retirement system for those employed in higher education), the percent of professors retiring under age 64 increased consistently between 1976 and 1985. Although not necessarily germane to the present paper, it is worth considering what such trends toward earlier retirement bode for the future. Let us, however, turn now to a consideration of the life course of that generation of Americans which came of age during the Great Depression.

PRIVILEGE BY ACCIDENT OF BIRTH: THE DESERVING RICH

To accord privilege of subsidy to a segment of society on no other grounds than the nature of their birth is reminiscent of medieval practices and seems inappropriate in a modern democratic society. Yet when part of the population is accorded rewards because of its age alone, then it is accident of birth which is the criterion. The date of one's birth is one such accident; the characteristics of the historical period through which one lives is a second accident of birth. Let us briefly review some of the history of the eras during which the two generations under consideration came of age. It is important to recognize that we are not addressing issues of deference and respect for the elderly, but rather the issues of financial subsidy based solely on age without regard to need. Although the caveat should not be necessary, we need to remember that a compassionate society is obliged to provide for whomever among its citizens are unable to provide for themselves--whatever age they may be.

A person who turned 60 in 1980 was born in 1920 and grew up during the Great Depression of the thirties--a time in America when money and jobs were scarce and poverty was endemic. At the time the sudden turn in social and economic events seemed a shocking thing. This is the generation of Americans we call the "depression generation" and it has never gotten over the trauma of those times or the fear that they could occur again. It has always been a generation of fabled ants, storing away reserves against some future catastrophe, avoiding debt, working hard and faithfully in preparaation for whatever needs might someday arise. Regardless of its current wealth this remains an economically conservative generation. Given the choice it will save rather than spend and it cautiously avoids taking risks with its capital. To this day, the depression generation irritates its YUPPIE children by constantly turning out electric lights, clipping coupons which save a few cents on groceries, and refusing to throw out any scrap of food which might still be edible.

America did not enter the second World War until the end of 1941 and nearly all of the young men of the depression generation found themselves drafted for military service as economic conditions began to improve. Those who were not in the military were either physically handicapped or were trained and skilled in ways which made them more useful as civilians than as riflemen. By 1946 the war in the Pacific had ended and most of these young men in their early twenties were returning to civilian life and to subsidies provided by a grateful and victorious nation. They not only received cash bonuses from both the federal and state governments and unemployment compensation for up to a year, but the GI bill provided among other things access to new houses and to education.

Many of this group went from high school diploma to Ph.D. or professional degrees with all tuition, fees, and books paid for as well as a monthly cash stipend. When they married, they were able to purchase a new home for little or no cash outlay and with government guarantees to the lenders. Our "baby boomers" were born and grew up under these prosperous conditions. In spite of a nasty little war in Korea and the emergence of the Cold War, jobs were plentiful, the economy was generally good, and the depression generation along with its future YUPPIE children prospered during the placid and plentiful fifties. Of considerable future importance was the fact that, as these people used the equity in their homes to purchase larger and more expensive housing, they were excused from taxation on whatever profits they made along the way.

This placidity began to evaporate in the sixties with the impact of the civil rights movement, the assassinations of its leader and of the nation's president, and the Vietnamese war. It was the children of the prosperous middle class survivors of the Great Depression and the "Last Good War" who rebelled against this new war, against the educational system, and against their parents' values in general. These were the "flower children" and the generation which tried everything--including drugs. It was the generation whose slogan was "never trust anyone over thirty." It was the first generation to grow up with the constant companionship of television and the constant threat of nuclear annihilation. It was a generation whose experiences in coming of age had not been shared by its parents and therefore could not be fully understood by them. The reverse is also true. These prosperous, consuming, spending children could not grasp the depression era experiences which resulted in a saving, conserving, deferred gratification world view among their elders.

As the decade of the seventies came to a close, peculiar things were happening to the American economy. Housing values began increasing at enormous rates and interest rates rose to the high

teens.   The depression generation began to retire from the labor
force in the eighties, some of them under age sixty and many of
them under sixty-five--the traditional retirement age.   Although it
was a time of economic uneasiness with both inflation and unemployment rising rapidly, the depression generation was able to
liquidate the houses it parlayed from little or nothing, for huge
profits, frequently in six figures.   And in the end, if they were
older than 55, they were excused from taxes on the bulk of these
profits.   The tax free money they received for their inflated homes
could be invested conservatively, frequently with government insurance, at rates as high as 16-18%.   Thus, while the American
economy was in a generally troubled state, the upper-middle class
depression generation increased its wealth by leaps and bounds.

Its YUPPIE children meanwhile were struggling to find jobs
and to find the capital for the kinds of housing and investments
which they had seen their parents enjoy.   Although some of them
had family incomes in six figures--far in excess of what their
parents had earned, they found it difficult to save money and
sometimes roused the contempt of their elders who blamed the consuming mentality of their children for their failure to amass capital.
The older generation having disposed of the suburban home or the
rural farm for immense amounts of money tended to fault their
"grasshopper" children for their inability to amass capital or to
make a go of it as farmers.   Recall, that no sooner had the elderly
farmers disposed of their land at vastly inflated prices, then the
value of farm land collapsed in America, leaving the young pur-
chasers holding the bag.

The older generation tends to see itself as a "deserving rich"
who credit their success to individual ingenuity, prudence, deferred
gratification, and perseverance.   They may not recall some of the
subsidies they received during their lifetimes and, by and large, see
themselves as deserving the subsidies from both public and private
sources which they continue to receive by right of age.   These
range from small everyday matters such as discounts on public
transportation, motion picture admissions, restaurants, hotels, and
liquor to reductions in larger matters such as real estate taxes and
the exemption of annuity and pension income from taxation in many
states.   The federal government has been the most generous of all
in its use of the income tax system as an income transfer system,
not necessarily to the needy, but certainly to the elderly.   Most
obvious is the annual double exemption allowed to persons age 65
and over, a priviledge modified but not eliminated for 1987 tax year.
Even more profitable is the one time exclusion referred to above, of
up to $125,000 in profit on the sale of a home.   These elders have
little sympathy for a young middle-class which tends to spend its
large income as fast as it earns it, and sometimes faster, on
whatever is chic at the moment: German cars, Scandinavian ice
cream, bottled water, or Caribbean vacations.   The depression

generation surely views such frenzied conspicuous consumption with self-righteous disdain. But if the well-to-do elders are disdainful of the life style of their youngers, it is possible that the YUPPIE view of the depression survivors is also less than generous. Are these not rich, old, selfish people who have already been given so much, who lived through such good times, who have so much, yet continue to demand privilege? Are they sometimes seen as miserly people clutching their wealth and unable to get pleasure out of the remaining years of their lives?

A modern version of Aesop's fable of the grasshopper and the ant, based not on scholarly historiography but largely on personal recollection and conjecture with whatever distortions those may entail has been presented. There was no reference to Max Weber or Thorstein Veblen whose concepts of the "Protestant Ethic" and of "conspicuous consumption" provide the foundation for this analysis. Surely, there are many people in both generations who do not fit the fable just spun. It is, however, possible that a condition of mutual generational contempt and distrust may emerge among the prosperous middle-class of America. The children who vowed never to trust anyone over 30 are now themselves in their late thirties and the generation they distrusted then is now in their sixties. What future implications does such a relationship hold?

Up to now we have considered some statistical evidence which is largely ambiguous and some historical ruminations which may be of questionable validity. The third section considers the types of imagery and perceptions of these two generations as purveyed in the media. This is perhaps the most important evidence since individual's actions are based on the beliefs they hold. If we are to anticipate future policies then we must grasp present beliefs. Both history and statistics will inevitably be mobilized to support whichever set of beliefs dominates as a guide for action.

MIDDLE-CLASS INTER-GENERATIONAL PERCEPTIONS

Although it is the opinion of this author, that the data now under consideration are superior to other kinds, it is necessary to point out that the collection of these data have not been systematic. They represent the author's personal collection of articles from newspapers and magazines as well as those provided by friends and colleagues who are aware of my interest. This information is primarily on the range of views and imagery being purveyed, and is not information that would allow us to test which perspective is predominant. The materials fall into four categories which will each be treated in turn:

- "Elder Bashing" articles and editorials which warn of the imminent threat of a gerontocracy;

- The defenders who claim that the arguments of the first group are without foundation;
- The peacemakers who write that there is no real issue of intergenerational conflict;
- A very small set of columns and editorials which attempt to view the issues in a detached and reasonable manner.

The Elder Bashers

It is easy to disregard as the vituperate propaganda, which it is, some of the extreme right attacks on the elderly, usually pretending to be in defense of the young who must pay the bills and, it is argued, will never receive social security benefits of their own because of the manner in which they are now being squandered. Paula Schwed, for example, writes to college students about "A Dirty Little Secret: You Support A System That May Never Support You. Social Security Can Be Cut" (Campus Voice, the National College Magazine, Aug./Sept., 1986). Amid flowery language referring to "bleeding heart liberals" and the like, the article warns the children of the eighties that they can never expect to live in as comfortable a lifestyle as their parents.

In a much publicized speech in September, 1986, Colorado Governor Richard D. Lamm accused the elderly, who were no longer disproportionately poor, of benefiting at the expense of the young. As early as February of 1985 the President's Council of Economic Advisors issued a report claiming that the elderly were no longer disadvantaged and supporting the claim with evidence of the kind discussed in the statistical analysis above. In June of 1985, Phillip Longman writing in the Atlantic Monthly called for "Justice Between Generations," again pointing out that the elderly were no longer poor and warning the baby boomers that they were headed for a disastrous retirement. Citing a study released by the conservative American Enterprise Institute, Spencer Rich wrote in The Washington Post in November, 1986 of the increasing poverty among the young and less educated and the improving conditions of Black and single parent families.

Despite its generally liberal editorial policy, The Washington Post contained an editorial on "Pandering to the Elderly" in May of 1986. It was concerned about the unwillingness of Congress to deal "realistically" with the social security issue. That same month an elderly reader of Newsweek named Joseph King expressed his guilt about being able to afford more but having to pay less for many commodities. His most telling example was the poor students whose theater tickets cost twice as much as his senior citizen discounted ones. Both the Post Editorial and the Newsweek articles brought in a rash of angry letters attacking the writers and defending the elderly. Let us turn from the bashers to the defenders.

The Defenders of the Elderly

In January of 1987, The Washington Post reprinted a number of "findings" from a report published by the Villers Foundation. The report decried and denied some of the elder bashing described above and the Post editorialized that "Poverty is a daily fact of life for several million elderly Americans." The same Post writer who had reported the conservative Heritage Foundation data now provided equal time to this new report which, among other things, defended social security as a sound program. Even a small neighborhood newspaper for elderly citizens in a suburban Kansas City county argued forcefully against what is called "The Myth of Elderly Affluence." A recent scholarly paper by a sociologist concludes that the "recent crises in Social Security and Medicare are in a large part manufactured by those who are ideologically committed to discrediting the liberal consensus of the past fifty years" (Hess, 1984). In addition to the bashers and the defenders, one finds a third group of writers who deny that there is any real issue at all.

The Peacemakers

This group reflects an awareness of potential for intergenerational differences, and some concern, but holds the position that with a degree of reasonable planning no conflict is necessary or likely. The leading voice of this group is the largest American organization for the elderly, The American Association for Retired Persons (AARP). They are displeased with the bashers. We say 'Enough!' to those trying to drive a wedge of greed between generations," writes Cyril F. Brickfield, Executive Director of AARP (news Bulletin, Oct., 1986). AARP, however, appears to hold a reasonable and understanding position. They recognize that there are other groups of needy in America and make a point of their sympathy for poor children. They simply refuse to accept responsibility for federal budget deficits or the role of "selfish narrow minded older people who care only about themselves."

A lengthy and carefully reasoned discussion of the issue appears in AARP's slick bi-monthly magazine under the title "The Phony War: Exploding the Myth of the Generational Conflict Between Young and Old" (Modern Maturity, Feb/March, 1987). Elliot Carlson, the author of that piece reviewed a considerable amount of scientific literature and interviewed a variety of leading social scientists. It reflects an understanding of the problem and issues and the concerns of the bashers. AARP does not deny that future retirees will get less of a return on their social security contributions than do present ones; it does, however, deny that the system is somehow corrupt and economically bankrupt.

A relatively new organization, Americans for Generational Equity, can tentatively be placed in this group. It is led by elected

and appointed politicians supported by a coterie of somewhat conservative scholars. It claims to be an advocate of the younger
generation and appears mainly concerned about the federal deficit
and the social security system. In his review of the AGE conference in January, 1987, Alan L. Otten writes in the Wall Street
Journal that "they dwell particularly on the need to overhaul Social
Security and Medicare now to make them more secure for later retirees." The arguments and evidence of the AGE people seem no
more "elder-bashing" than the arguments of AARP are a greedy defense of self- interest.   Both organizations represent tenable
positions, and appear reasonable in their efforts to do so.
     In addition to their pet economic experts (economists are like
psychiatrists in that one can always find an "expert" to provide
scientific support for one's position), both groups appear committed
to traditional family values involving respect for the elderly and
intergenerational dependence. AGE, for example, cites Neugarten to
this effect and The Wall Street Journal report of a recent AGE
conference cites a volume published by the American Gerontological
Society, "Ties That Bind," which stresses generational interdependence. To the extent that organizations like AARP and AGE can
remain free from the influence of the more extreme elements which
are attracted to them, it is possible that reasonable people will be
able to strip away the propaganda and vested interests--to identify
real problems and thus move toward real solutions.

The Detached Commentators

     This group, small and unorganized, appears anticlimactic
compared to some we have been discussing. Yet because they have
no axe to grind--no partisan constituency--they may more accurately
reflect directions in which the society is moving, than do the par-
tisans. For example, Stanley Jacobson, a local psychologist, writes a
scholarly piece on stereotyping the old, in the Washington Post
(March 1, 1987). He argues that a chronological age category such
as 65 and older makes no sense at all and suggests that AARP itself
may promulgate stereotypes harmful to the elderly. He discusses a
considerable list of elderly heroes with different life styles in their
older years, ranging from George Burns to Armand Hammer.
     Even more suggestive of popular sentiments is a piece written
by the Boston syndicated columnist, Ellen Goodman in late 1986.
She writes of "Rebalancing the Family Checkbook." She says that
occasionally something comes to our attention which suggests that
things are not as they ought to be. It may be the sudden awareness that our country neglects poor children while rewarding well-
to-do elders (a comparison which AARP rightly points out is a non-
sequitur), or noticing on the income tax form that everyone over
65 regardless of wealth gets an extra exemption, or you see the
large chunk extracted from your child's pitiful pay check for social

security tax.   With Goodman, the awareness hit when she read
Sheila Graham's Hollywood column, written from her luxurious home
in Palm Beach, in which Graham admitted that "it gives me satisfaction to pay half-fare on busses and trains and only $2 at the
movies."   Here is the personification of the depression mentality
discussed earlier in this paper.   Goodman's column reflects well
informed concern and a feeling that something is out of adjustment
and needs fixing.   This may be the direction in which public
sentiment is moving.

CONCLUSIONS AND DIRECTIONS

To a certain extent it appears that there is a concentration of
wealth in the upper-middle class segment of the American elderly
which considers itself deserving and which carries with it some
political power. The elderly in America are well organized and have
strong voices both in state legislatures and in the U.S. Congress.
The executive branch, not known for its generosity, has twice found
it prudent for the president to authorize cost of living increases for
social security recipients even when the law did not require such
increases (COLA's). Analysts such as myself do not intend to par-
take of elder bashing and feel uncomfortable in the company of
right-of-center ideologues.   But, in addition to a professional
curiosity, we are motivated by a sense of justice.   If, in fact, the
rich are being rewarded at the expense of the less rich, then
alterations are needed in the system--alterations which will do no
damage and may help those who are in economic need--old and
young, men and women.
The United States is rich enough and compassionate enough to
help all who have inadequate resources. Although a policy which
chooses to ignore age and sex categories seems just and reasonable,
it is hardly necessary, in our prosperous country, to make choices
among those citizens who need financial assistance. Social justice
demands that we find a means of helping old and young, men and
women, and that we find a means of doing so which is not
demeaning. In a rich country all citizens are entitled to adequate
financial resources.
It is difficult to get a handle on this issue, in part because
the data are so ambiguous. Age and income or age and wealth
analyses can be shaped to nearly any form the analyst wishes. And
there are the statistics of fear based on unreasonable assumptions
and designed to create apprehension that "we will all die poverty
stricken in nursing homes or the social security system will be
bankrupt in precisely X months." Fictional projections are no more
helpful than ambiguous data. Yet there are data that appear more
persuasive, such as demographic data that documents the existence
of a "baby boom" moving through the aging population pyramid or

the gradual increase in life expectancy. However, the existence of valid data does not guarantee valid interpretation. This is also true of historical data such as that presented in this paper. Different analysts would suggest that in fact different events occurred and with different consequences.

The only data which is known to be accurate in the sense that it is useful in determining the future course of events, is perceptual data. When people believe that something is true they act on that belief. Whether the belief can be demonstrated to be factually correct or not is irrelevant. As Herbert Blumer (1948) pointed out in his analysis of public opinion, not all people are equal in the impact their beliefs have. It is likely that in contemporary America the mass media and the people who talk to us through it are the central figures both in reflecting and shaping the beliefs that lead to actions. It is for this reason that columns and articles from newspapers and magazines appear to be so crucial a source of data. The contrast between the contemporary scene and some of the pioneering sociological analyses of intergenerational relations (e.g., Sussman, 1960; 1965; Sussman & Burchinal, 1965) demands new systematic analyses of the depression generation and their YUPPIE children.

Methodological Issues

This final section will address some methodological concerns and questions. For example, how does an analyst determine what is a reasonable sampling of the media? Is there a systematic way to analyze such data so that one's analysis will persuade others? Are these the most desirable kinds of data for understanding this issue? Of even greater concern is the difficulty in locating a hook to hang all of this on. Although personal observation suggested, at first, that new territory was discovered, a colleague-gerontologist quickly revealed this misperception by using the term "gerontocracy" to describe these concerns. A word like that reflects some sort of prior interest. Furthermore, the literature contained materials on so called "intergenerational relations." Apparently the ideologues had arrived before me.

The question is what can be done that others are not doing or that needs to be done in a different way?

▒ One can monitor public images and watch for signs of change;

▒ One can focus on benefits other than social security and medicare which appear to receive most of the attention, e.g., issues such as tax policies or private and commercial benefits;

✳ It might be reasonable to concentrate exclusively on the upper-middle class--the depression generation, and their YUPPIE offspring;

✳ Finally one might search for a link between generational relations and the type of economy and society in which they occur.

POSTSCRIPT: TOWARD A COMPARATIVE ANALYSIS

In the tradition of de Toqueville, who thought well of things American, and Dickins who did not, a clearer perspective on this issue might emerge by distancing the analysis from its domestic setting. One might consider the unlikelihood of generational relations of this type developing among third-world nations which do not have a large well-established middle class. Is such a relationship possible among socialist countries which attempt to provide relatively equal resources to their populations and whose public policies discourage the amassing of wealth? What of the true Welfare State in which tax policies are designed to prevent the accumulation and retention of wealth? Perhaps the conditions we describe can exist only in industrialized capitalist nations such as the U.S.A. or West Germany or Switzerland. Yet the "pension elite" in the Soviet Union are mentioned in the literature and both British and Dutch colleagues recognize the phenomenon considered in this paper.

Comparative work might require other varieties of methods since not all societies tolerate an uncontrolled, gossipy press to serve as a data source, or permit the analyses of age, income, or population data from whatever perspective the analyst wishes. In fact, not all societies have such data. Yet, it does seem that efforts at cross national analysis of the historical roots, family traditions, and current conditions would illuminate the situation and provide a better understanding of the economic images the generations have of one another and their implications for the society at large.

In conclusion, if it is possible for a powerful minority to impose its will on a less powerful segment of the population, insisting, like kings of yore, that the less able be taxed to support the well-to-do, this poses a serious threat to democratic society. This threat transcends generational conflict and moves toward an elitist society based on the sole criterion of accident of birth. Who is the underdog under those conditions? ✳

A version of this paper was presented at the Annual Meeting of the Society for the Study of Social Problems, Chicago, 1987. The author is indebted to Professors Warren A. Peterson and Robert W. Habenstein along with the trainees of the Midwest Council for Social Research on Aging for their incisive critisisms of earlier drafts.

ENDNOTES

1. For a sociological analysis of this issue focusing exclusively on Social Security and Medicare, see Hess (1985).

2. The terms "relatively rich" and "relatively poor" are mine, not Radner's. He would have preferred to compare the first and tenth deciles but found those two extremes flawed in ways which compelled him to use the less extreme ones (Radner, p. 67, n. 10).

REFERENCES

Becker, H.S.   Whose side are we on?   *Social Problems,* 1961, 14, 239-47.

Blumer, H.   Public opinion and public opinion polling.   *American Sociological Review*, 1948, 13(5), 542-54.

Carlson, E.   The phony war.   *Modern Maturity*, 1987, 30, 1(Feb/ March), 34-36.

Gouldner, A.W.   Anti-minotaur: The myth of a value free sociology. *Social Problems*, 1962, 9, 199-213.

Grad, S.   Incomes of the aged and nonaged, 1950-1982.   *Social Security Bulletin*, 1984, 47, 6(June), 3-17.

Hess, B.   The withering away of the welfare state: Manufactured crises in policies for the aged.   Paper read at the annual meeting of the Eastern Sociological Society, Philadelphia, March, 1985.

Longman, P.   Justice between generations.   *The Atlantic Monthly*, 1985, June, 73-81.

Radner, D.B.   Changes in the money income of the aged and nonaged, 1967-1983.   *Studies of Income Distribution.* 14(Sept.), U.S.   Department of Health and Human Services, Social Security Administration.   Washington, D.C.: U.S. Government Printing Office, 1986.

Schultz, J.H.   *The Economics of Aging.* (2nd Edition).   Belmont, CA: Wadsworth, 1980.

Streib, G.F.   Social stratification and aging.   In R.H. Binstock & E.S. Shanas (Eds.), *Handbook of aging and the social sciences* (2nd Edition).   New York: Van Nostrand Reinhold Co., 1985.

▓ 10 ▓

TRACING THE DISADVANTAGES OF FIRST-GENERATION COLLEGE STUDENTS: AN APPLICATION OF SUSSMAN'S OPTION SEQUENCE MODEL

Margaret Brooks-Terry

INTRODUCTION

Since the term "first-generation college student" entered the vocabulary of higher education specialists within the last decade, the concern has centered on the performance of these students in the academic setting. Research has shown that college students whose parents have no personal experience with university-level education have more problems in social adjustment to college, are less likely to be involved in campus organizations, and are more likely to drop out before graduation than are second-generation students. Recent programs designed to aid students in the adjustment to college [(e.g., the Freshman Year Experience programs (Gardner *et al.*, 1987)] have been structured in part to address these problems. Little attention has been paid, however, to the social processes by which first generation college students are disadvantaged.

The following analysis borrows from the option sequence model advanced by Sussman (1972) to explain actions taken by individuals entering retirement. Elements of that option sequence model are employed to explain the family processes that result in adjustment problems for first-generation students in the college milieu. Like new retirees, students entering college are acting on choices made under some degree of uncertainty. Each is either extending or re-entering the status of dependency and subordination. And the options available, in principle, to all persons who reach the appropriate life stage are, for both the high school graduate and the older person leaving the work force, constrained by characteristics of their subcultures, their families and by their own personal qualities.

FIRST GENERATION COLLEGE STUDENTS

First-generation students are defined as those whose parents have not had a college education. Parental education is, of course, one of the principle determinants of the family's position in the social class structure. But education and social class are not synonymous. The zero order correlation between education and occupation, and, between education and income for American males, is about .60 (Blau & Duncan, 1967; Coleman & Rainwater, 1978; Jencks *et al.*, 1972). At the level of community and peer groups, first-generation college students are influenced by the family's social class. However, socialization at the most fundamental level, within the family itself, is a product of the specific values, attitudes and behaviors of a particular set of parents. Parents communicate expectations and life goals to their children based on their own educational experiences.

Second-generation students come from homes in which parents have experienced the university environment, although in some cases the parents have not completed degrees. The principle defining characteristic of second-generation students, however, is that their parents--the generation whose influence is greatest through the developing years--have personal knowledge of higher education (Billson & Brooks-Terry, 1982). Research on American college students has identified several characteristics differentiating first-generation from second-generation students.

Higher attrition rates. It has long been recognized that first-generation students are overrepresented among students who leave their first college or university, and who leave higher education for good--particularly during or just after their first year. Only a small proportion leave because of academic failure (Stanfiel, 1973; Astin, 1975).

Conflicting loyalties. First-generation students are more likely to live at home and to work part-time at a location outside campus. The family and the work setting each represent a set of values contradictory, in part, to the values of the university. Daily participation in the role systems of each of these social settings results in conflicting demands for the time, attention and energy of the students (Billson & Brooks-Terry, 1982; Brooks-Terry & Billson, 1987).

Pragmatic goals. For first-generation, even more than for second-generation students, higher education is perceived as a means to a secure and well-paying career. Job-related skills are the valued outcomes of college courses, and job-specific majors, such as accounting and criminal justice, are the most frequently selected academic programs (Billson & Brooks-Terry, 1982).

Lack of involvement. First-generation students tend to be less involved in campus organizations and activities. Living at home with parents and holding down a part-time job means less time to devote to non-essential activities on the campus. The result is not only weaker ties to the educational institution but fewer close friends on the campus (Billson & Brooks-Terry, 1982). Astin (1984: 299) demonstrates that the amount of involvement, which he defines as *"physical and psychological energy that the student devotes to the academic experience,"* is directly related to the likelihood of persisting in college.

Double assignment. The assignment for all college students is to prove their mastery of the course material at a level high enough to graduate and enter a career or advanced educational program. For first-generation students there is a second assignment, to discover and internalize the lifestyle of the college-educated middle class. The burden of this double assignment is not generally recognized by the students and their families, nor by faculty and advisors in colleges. The value system of the family and community from which the first-generation students come may be fundamentally different from that of the university. In order to achieve his or her career goals, the student must reject the values of home, peers, and neighborhood, and take on the attitudes and behaviors associated with the work world he or she wishes to enter. Those most successful in preparing themselves for social mobility through education and work will do so at the cost of weakening their bonds to the family unit. Whether consciously or unconsciously, some will be unwilling to make that sacrifice (Brooks-Terry & Billson, 1980). By contrast, second-generation college students are treading in the footsteps of their parents. They come from backgrounds that espouse values similar to those of the university community. These students are comfortable in the college setting since their parents have been preparing them for college since childhood. Graduation affirms their place beside their parents in the white-collar middle class.

Doing college on the side. Students who live with their families and work off-campus, who see college as simply a means to a desirable job and are not involved in the total college environment, are unable to give education their full attention. In the words of one student:

> *I sometimes miss classes because the manager changes*
> *my work hours or my mother needs my help at home.*
> *I feel as though I am doing college with my left*
> *hand while my real work is elsewhere.*

These students leave campus daily as soon as their classes end, assert that they have no time to complete lengthy library assignments, cannot attend evening functions on campus, and generally appear (and are) marginal to the life of the university (Billson & Brooks-Terry, 1987). Those who persevere to complete their degrees are likely to be disappointed in the career opportunities available to them, in part because they have not become comfortable with the norms of the work world they aspire to enter.

What are the social processes that result in different student experiences? How is it that families, all of whom want the best for their sons and daughters, can unwittingly create the obstacles that often frustrate or abort those goals? The search for answers to these questions leads us to a model of the family as linkage and limiter between cultural ideology and societal options.

THE OPTIONS SEQUENCE MODEL

Sussman identified sets of factors impacting the selection of options by those facing retirement. Four levels of the model also determine the options available to the high school graduate: outer boundaries at the societal level, situational and structural constraints, the family as a linkage system, and individual variables (Sussman, 1972). The component parts of the influence model allow us to compare first-generation and second-generation students who have made the initial decision to attend college. At the societal level we begin with the cultural values that unite Americans across educational and economic levels. Second, we move to the specification of these global values in different subcultures. Third, we consider the role of the family as a socializing agent and intergenerational linkage between its members and the institutions of the society. Finally, there are the individual variables which enhance or override the influences of the other components (See Figure 1).

The Societal Level: Outer Boundaries

Sussman (1972:48) describes the outer boundaries of the model as *"circumscribing the limits within which options are exercised in relation to careers in various life sectors."* Democracy, individualism, independence, and work and equal opportunity for all, are among the bedrock values of American culture. Adult status is defined by financial independence, work, and adult social responsibilities; dependence is humiliating. Success is believed to be earned on the basis of individual merit. The active expression of these cultural values is the open access to education and the weight given to achieved status in the United States. Americans across all social strata accept these values and believe that, in general, the options are available to all who are willing to invest sufficient effort.

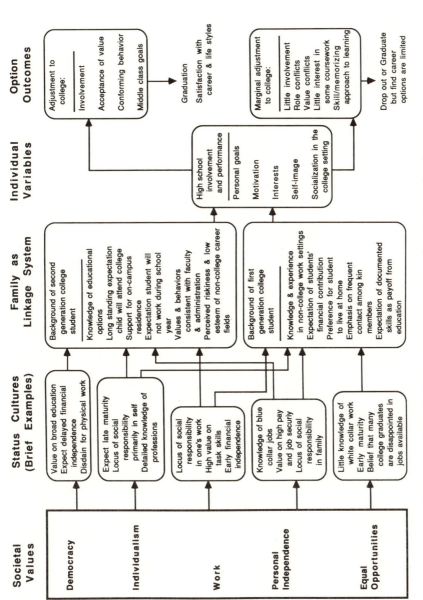

Figure 1. The Options Sequence Model for College Students

The faith placed in such terms as "independence" and "work" masks the differences in meaning given to these qualities across the social strata. These class-based differences, or status cultures, are reinforced by interaction among status members. For example, in the intellectual context of one status culture the notion of individualism means defending oneself against assaults from peers. In another it could include defiance of parental wishes and perhaps even the choice of an alternative lifestyle. The definition of work in some settings presumes the earning of income; in others it might include the work of pursuing extended education or years of dedication to the Peace Corps.

Generational transmission is the process through which this status culture is conveyed from one generation to another with the result that each social stratum maintains its intellectual context over time (Peterson & Rollins, 1987). The family of orientation is the primary agent of intergenerational socialization.

Education plays a dual role in determining social status. First, educational settings themselves socialize participants with the values necessary for each status culture to replicate itself. Public high schools impart the conformity and obedience required for working class jobs with high levels of standardization and little autonomy. College socializes students for self-direction and thinking in terms of relationships as called for by managerial and professional careers (Langman, 1987; VanFossen, 1979). Thus the education of parents helps to establish what Blau and Duncan (1967) call the *"educational climate"* of the family.

Second, education is the primary avenue of social mobility in contemporary American society. The first-generation college student, who is in the process of being socially mobile, must successfully complete the attitude and behavior transformation to middle class characteristics. Achieving consistency between the status conferred by his or her education and characteristics of personal behavior is the double assignment the first-generation student carries during the college years.

Status culture influence on options. Status and ethnic cultures limit the options available to their members in a variety of ways (See Figure 1).

Definition of developmental stages. The definition of adult behavior, especially as it pertains to financial independence, and the age at which it is expected, plays an important role in the support and encouragement of continued education. In some cultures, the eighteen year old is expected to support himself, at least in part. In others, maturity is conferred by the family at the same time the final degree is conferred by the university.

▓ *Myths and values.* Children who grow up in homogenous communities are likely to be embedded in a strong peer group whose values reflect those of the parents. People have a better understanding of the environments they have experienced. Part of every status culture is the shared myths about other lifestyles. Therefore, knowledge of career options and how to achieve them is likely to be as limited for adolescents in the middle class as it is for those in the working class, but the opportunities they know about are different (VanFossen, 1979).

▓ *Values regarding productivity.* "Work," "earning a living," and "education" take on different meanings in different subcultures. In some value systems the weight accorded to "paying one's own way" is greater than that attached to a college degree. In others, the weights are reversed.

▓ *Social responsibility.* Some status cultures define the family as the primary locus of loyalty. In other cultures one's work takes precedence. And, in still others the focus on individualism is carried through to emphasize loyalty first to one's own goals. Individual priorities of time and energy are distributed according to these loyalties.

The Family As Linkage Agent and Facilitator

Families link the status culture across generations. In addition, parents reflect their own life experiences in the socialization of their children, modifying the subcultures as they perpetuate them (See Figure 1). Parents' own exposure to college is a principle factor in their sons' and daughters' perception of college as an option, selection of a college to attend, and adjustment to the college environment. In fact, Sewell and Hauser (1975) concluded that parents and peers have about equal influence on boys' decisions to attend college, and both are two to three times as influential as teachers and counselors.

Blau and Duncan (1967), Jencks (1979), and others have examined the influence of parents' own education on the level of education completed by their children, but their work says little about the dynamics of parental influence on the process of their children's education. Bowles and Gintis (1976), however, suggest that it is the attitudes, values and behaviors acquired in the process of higher education, rather than the knowledge of particular fields of study, that are the critical payoffs from college. It is these attitudes and lifestyles that separate parents who have attended college from those who have not.

Family influence on options.   There are seven ways in which the family can limit or facilitate the students receptiveness to options.

🟥 *Feasibility of higher education.* Many parents who have not had the opportunity to attend college themselves want their sons and daughters to benefit from higher education. Concerns communicated by mass media and popular mythology, however, appear as major obstacles: cost, failure to find employment at graduation, fear of failure. College educated parents tend to communicate throughout their children's developing years the expectation that they will attend and succeed in college. As one second-generation student responded when asked how the decision to attend college was made: *"That was never a topic of discussion. It was always assumed."* First-generation students are much more conscious of having considered the options of university versus immediate employment (Billson & Brooks-Terry, 1985).

🟥 *Selection of the institution.* While many students believe that they have chosen the college they enter, in fact, the selection was generally made from among a small set of alternatives approved by the parents. The basis upon which parents decide which colleges should be considered is a function of their beliefs about those institutions (VanFossen, 1979).

🟥 *Residence.* For some students, living at home facilitates options because it makes college affordable. But, there are other reasons for living at home as well. Parents worry about the stories of alcohol and drug-use, and about sexual promiscuity. Frequently, their well-socialized sons and daughters share those worries: *"I'm afraid to live in the residence hall,"* said one entering freshman when asked where she would be living. Students who live at home are limited in opportunities for involvement on campus, a particular disadvantage for first-generation students.

Parents who have attended college have values and behaviors generally compatible with those of the college community. They are more likely to view the residence hall experience as a valuable part of college life and define occasional behavioral excesses as immaturity--tolerable, if not desirable.

🟥 *Expectations of coursework.* Observing the amount of free time many high school students have, and the brief, content prescribed assignments they work on, it is no wonder that many parents assume their sons and daughters can hold part-time employment while enrolled in a full-time academic program. The fact that working ones way through school was probably always more myth than reality is unknown to them. Parents who have attended a university are more likely to understand the uneven pacing of assignments throughout the academic term, recognizing that laboratory research, lengthy papers, and major examinations do not fit well with the routine schedules of most off-campus employment or care of younger siblings. If the student is a resident on campus, and does not work during the academic year, the college is a total institutional setting in which the expectations are consistent and mold the student in a single direction.

* *Competing role sets.* The college student living with parents experiences a daily reinforcement of the subculture in which he or she grew up. To the extent that the values and behaviors of the family environment differ from those of the college milieu, the student will feel the stress of dual loyalties on a daily basis. When there is an off-campus work setting, a third subculture, complete with role relationships and claims on loyalty, is a daily component of the student's life. Leaving home in the morning, attending classes into the afternoon, and going directly to a part-time job means a three-cornered journey through conflicting social worlds.

* *Knowledge of the higher education system.* When to apply, what counts on applications, sources of financial aid information, college resources for tutoring, recourse for grading disputes, and all the other aspects of college life are familiar to those who have used them. Without personal experience with these institutions, neither parents nor students may know the options that exist, and parents may offer inappropriate advice. Most college advisors have encountered the parents who insist that their son *"does not need to take English 101 because he is going to be an accountant."*

* *Definition of the "unknown."* Higher education is a costly investment. Parents who are confident of the outcome will support the decision. Those who have heard that some college graduates are unable to find employment based on their college degree fear that time and money will be wasted (Collison, 1987). For those with college backgrounds, it is the lack of education and the career options associated with that status that are unknown and appear risky.

Individual Variables

The way in which the individual handles the constraints described above mediates their influence. Five individual variables are especially important among these personal inputs (See Figure 1).

High school performance. Long known as one of the best predictors of college performance, high school grades reflect an attitude toward education and study skills, as well as knowledge. Good high school students tend to have internalized the value of education, no matter what their background (Sewell, Haller, & Ohlendorf, 1970).

Personal goals. The first-generation student who sees himself as a chemist--with some idea of what a chemist is and does--is likely to persist in his studies. He or she will seek role models among the chemistry faculty who substitute, to some degree, for the loss of identity with kin and community of origin.

Motivation. Goal orientation and motivation tend to be correlated, but when motivation is only focused on the degree as a credential, there may be an insufficient change in values and behavior to assure the intended social mobility.

Interests. The student who brings to college one or more interests that bond him to other students in the classroom or in extra-curricular activities ensures an opportunity for socialization through involvement.

Self-image. Confidence and a personal sense of independence from the family, free the individual to loosen old ties.

CONSEQUENCES OF THE OPTION SEQUENCE

Sussman (1972:67) described the option sequence as "*a funnel-shaped process*" in which the options that are recognized and believed to be available are limited by social circumstances. Across social strata, Americans subscribe to a belief in equality of opportunity and merit-based access to resources. But in reality, different life situations limit the options available to individuals. In the case of first-generation college students, the selecting out begins with the level of option awareness which is constrained by the knowledge and values of the subculture.

In communities with generally high educational levels, high school students know the options of higher education in some detail. They generally participate in a status culture that shares knowledge of higher education resources informally, expect prolonged dependency to accommodate extended education, and value a broad base of knowledge. The social network of family friends can be tapped for knowledge of fields open to those with extensive education, and there is confidence that those careers are attainable and satisfying. To a significant degree, the status culture itself chooses college for the sons and daughters of the highly educated middle class. It is the option of conformity.

Adolescents from blue collar environments generally develop their values and expectations in a very different status culture. The options associated with early financial independence and immediate entrance into the labor force are the "knowns." Physically demanding work is often an important component of masculine identity, as are home-oriented roles for women. The varieties of higher education, the life of the student, values taken for granted by faculty, and the kinds of careers at the end of the line are all uncertainties of varying degree. To opt for college may bring esteem from the community, but little useful information or support. The young adult's journey to college begins with a break from the status culture and increases the social distance from home and community as the education career progresses. It is the nonconforming option.

No matter what the community of residence, once the decision to attend college has been made, it is the parents' educational experience that structures many of the options available to the student. Before the student ever enters the college classroom a number of choices have to be made. Students may gather information about colleges, but it is parents who usually set limits of cost and distance from home, and sometimes even the majors that can be considered. First-generation students generally select from a set of college options that are close to home, lower in cost, and promote a number of majors with direct occupational outcomes. Parents often encourage the student to live at home on the premise that board and room are unnecessary expenses. But, commuting to classes usually necessitates a car, and the costs associated with transportation lead to part-time employment. Living at home results in expectations about household chores, kinship social obligations, and accountability for time that further limit the options for campus involvement. Even when first-generation students live on campus they tend to go home on week-ends or whenever there is any kind of family crisis. The obligations to family and work make the first-generation student a marginal member of the campus from the day he or she enters classes.

If the student masters the "double assignment" of value and behavior changes consistent with the career goal, social distance and friction with family members is inevitable. There is an element of paradox for the union-member parent who is proud that his son will be a manager and look like a manager, but is resentful when the son thinks like a manager and behaves like a manager. The parents' response to the changing subculture of reference influences the student's perception of career options.

The daily reinforcement of the expectations of parents, the ties to old friends in the community who are now earning regular paychecks, and the values demanded by the part-time job all exert strong pulls on the student to remain immersed in the old status culture. There are familiar, positively sanctioned options in these social settings that compete directly with the options associated with higher education.

The student's own commitment to the successful completion of higher education is a crucial factor in moderating the impact of the status culture and the parents' education. Sussman points out that commitment depends on satisfaction with the activity, the belief in a payoff, and the behavior of significant others. Individual students who have a very clear image of a career goal are often able to overcome these bonds to social structures outside college. Equally important, the highly motivated "good student," easily convinces parents that the investment in college is worth-while because "success" in terms of a well-paid, prestigious job seems assured.

But, the first-generation student who lacks a clear goal, who is not committed to a particular field of study, and who doubts his or her own capabilities and objectives, is vulnerable to the pulls of his other role sets. Dropping out of college is a salient option for such students. For those who remain in college, the double assignment may not be completed. The student is trained but not resocialalized. The result is the accounting major who is passed over by the major accounting firms, not because of poor grades, but because he or she does not fit the image of a corporate auditor.

For the second-generation student, going to college, accepting its values, and remaining to graduate minimize the social distance between the student and parents. The young adult's place in the status culture and in the family is reaffirmed by college graduation.

SUMMARY

First-generation college students, those whose parents have not attended college, tend to have lower levels of involvement in their educational experience than do second-generation students. First-generation students have problems adjusting to college because they bring with them values and behaviors that are incompatible with the cultural milieu of higher education. The option sequence model provides a framework for tracing the perception of choices available to the student and the way in which the options selected result in marginality.

The status cultures of class and ethnicity, together with the educational experience of the parents, are like a set of lenses. They focus on an option in clear detail or blur it beyond recognition, bring it closer, or remove it to an unreachable distance, or distort the option so as to create inappropriate expectations or

fears.   In addition, first-generation students experience multiple
forces pulling them away from the college setting.   Yet all of these
qualities that differentiate first-generation from second-generation
students tend to be overlooked by institutions of higher education or
even hidden in an effort to make sure that all students are treated
equally.   In the absence of very strong personal goals and a
willingness to sever some of the bonds with community and family,
it is no wonder that first-generation students have a higher college
drop-out rate. ▨

The study of first-generation college students, on which this paper
is based, was conducted jointly with Janet Mancini Billson of Rhode
Island College.

REFERENCES

Astin, A.W.   *Preventing students from dropping out.*   San Francisco:
Jossey-Bass, 1975.

Astin, A.W.   Student involvement: A developmental theory for higher
education.   *Journal of College Student Personnel,* 1984, 25, 297-
308.

Billson, J.M., & Brooks-Terry, M.   In search of the silken purse:
Career orientation vs. liberal arts orientation among first generation college students. *College and University,* 1985, 8, 57-75.

Billson, J.M., & Brooks-Terry, M.   Clientele for a changing era in
higher education: Retention strategies for first-generation students.
Paper presented at the New England Conference on Educational
and Occupational Counseling of Adults. Shrewsbury, MA, 1985.

Billson, J.M.   A student retention model for higher education.
College and University, 1987, 2, 290-305.

Blau, P.M., & Duncan, O.D.   *The American occupational structure.*
New York: John Wiley & Sons, 1987.

Bowles, S., & Gintis, H. *Schooling in capitalist America: Educational
reform and the contradictions of economic life.*   New York: Basic
Books, 1976.

Brooks-Terry, M., & Billson, J.M.   The double assignment: Academic
and social expectations placed on first-generation college students.   Paper presented at the Annual Meeting of North Central
Sociological Association, Dayton, OH, 1980.

Brooks-Terry, M.   Adjustment to higher education: A research
comparison of British and American students.   Paper presented at
the Second International Conference on the First Year Experience,
Southampton, England, 1987.

Collison, M.N-K. More young Black men choosing not to go to college. *Chronicle of Higher Education* 1987, Vol. XXXIV, 15, 1, 26-27.

Coleman, R.P.,& Rainwater, L. *Social standing in America: New dimensions of social class.* New York: Basic Books, 1978.

Gardner, John N., *et al.* Presentations at the Freshman Year Experience Conferences (1985-87). University of South Carolina, Columbia, SC, 1987.

Jencks, C. *Who gets ahead? The determinants of economic success in America.* New York: Basic Books, 1979.

Jencks, C., Smith, M., Acland, H., Bane, M.J., Cohen, D., Gintis, H., Heyns, B., & Michelson, S. *Inequality: A reassessment of the effect of family and schooling in America.* New York: Basic Books, 1972.

Langman, L. Social stratification. In M.B. Sussman & S.K. Steinmetz (Eds.), *Handbook of marriage and the family.* New York: Plenum Press, 1987.

Peterson, G.W., & Rollins, B.C. Parent-child socialization. In M.B. Sussman & S.K. Steinmetz (Eds.), *Handbook of marriage and the family.* New York: Plenum Press, 1987.

Sewell, W. H., & Hauser, R.M. *Education, occupation and earnings.* New York: Academic Press, 1975.

Sewell, W. H., Haller, A.P., & Ohlendorf, G.W. The educational and early occupational status achievement process. *American Sociological Review,* 1970, 35, 1014-1027.

Stanfiel, J.D. Socioeconomic status as related to aptitude, attrition, and achievement of college students. *Sociology of Education,* 1973, 46, 480-488.

Sussman, M.B. An analytic model for the sociological study of retirement. In F.M. Carp (Ed.), *Retirement.* New York: Human Sciences Press, 1972.

Sussman, M.B. The family today: Is it an endangered species? *Children Today,* 1978, 48, 32-37.

Sussman, M.B. Family relations, supports, and the aged. In A.M. Hoffman (Ed.), *The daily needs and interests of older people* (2nd edition). Springfield, IL: Charles C. Thomas Publisher, 1983.

VanFossen, B.E. *The structure of social inequality.* Boston: Little, Brown and Company, 1979.

▨ 11 ▨

EXCHANGE AND POWER IN MARRIAGE IN CULTURAL CONTEXT: BOMBAY AND MINNEAPOLIS COMPARISONS

Murray A. Straus

INTRODUCTION

The balance of power between husband and wife is one of the most important dimensions differentiating the family system of one society from another, and for differentiating families from each other within a society (Olson & Cromwell, 1975; Straus, 1964b; Szinovacz, 1987). At the same time it is an issue fraught with both methodological and theoretical controversy. Among the theoretical controversies are the extent to which societal norms determine the actual power relationships between spouses (i.e., a cultural norms theory of power) as compared to the extent to which power is the outcome of negotiation between husbands and wives on the basis of the resources and alternatives that each has available (i.e., an exchange theory of power).[1]

Among the methodological controversies are numerous questions about the validity of different techniques for measuring power. If the measure is based on interviews or questionnaires, can a typical respondent be expected to perceive and accurately report such things as *"who has the final say"* in respect to certain decisions? And what if one gets a different set of information from interviewing husbands compared to a wife or a child (Price-Bonham, 1976; Safilios-Rothschild, 1970; Szinovacz, 1987)? If the measure is based on observation of family members interacting with each other in their home or in a laboratory, do the circumstances distort their behavior? Perhaps most problematic of all is the question of whether the same research procedures used in societies as different as India and the USA produce equivalent data (Buehler, Weigert, & Thomas, 1974; Lee, 1982; Straus, 1969; Sussman & Cogswell, 1972). These are large and complicated issues which will not be settled by one study. However, the research to be reported in this chapter provides at least some data on each of the issues just listed.

RESOURCES AND CULTURAL CONTEXT

The "resource theory" of marital power (Blood & Wolfe, 1960: Bahr, Bowerman, & Gecas, 1974; Heer, 1963) is a special case of exchange theory. The theory asserts that the degree to which spouse A is able to influence the behavior of spouse B is a function of B's perception of the potential or actual ability of A to satisfy his or her wants. To the extent that A can or does provide "resources" which are important to B, and to the extent that alternatives are not available to B, A will be able to induce compliance to his or her wishes. However, the operation of exchange principles such as those implicit in the resource theory depends on the way the culture in which a couple lives defines appropriate roles for husbands and wives and also on the cultural definitions of what constitutes an appropriate resource.

The resource theory and its modification to account for the way in which each culture defines the terms of exchange (the "resources in cultural context" theory of Rodman, 1972), will be tested by determining the correlation of marital power with three types of resources: economic resources (such as occupation), interpersonal skills (such as creativity), and family structural resources (such as the presence of consanguineal kin). Each of these types of resources will be considered for the husband and wife and for the degree to which one exceeds the other ("relative resources").

The combined effect on power of all resources together will also be estimated by multiple correlations. The first hypothesis to be tested is that in both societies, the greater the relative resources, the greater the power. However, a second hypothesis holds that, with the exception of the presence of consanguineal kin, the correlations will be lower in Bombay. This is because Indian society--even in a metropolis such as Bombay--does not recognize the legitimacy of individual achievement to the extent recognized by European industrial societies. Instead, primary emphasis is placed on power derived from one's position in a web of kinship. These hypotheses will be tested using bivariate correlations for the individual resources and multiple correlations to estimate the combined effect of the entire set of resource variables.

METHODOLOGY

Sample

A questionnaire was given to children in schools selected on the basis of their location in middle class or lower class areas of Minneapolis and Bombay. This questionnaire permitted the selection of a stratified random sub-sample of 64 families which met certain requirements of a larger study (Straus, 1970). The resulting sample in each of these two cities included 32 families with male children and 32 families with female children. The children in the Minneapolis study were in the ninth grade; those in the Bombay study were in the seventh standard in schools in Marathi. In Minneapolis, the testing took place in a laboratory at the University of Minnesota. The testing in Bombay was conducted in a school or recreation center in the area where the family lived. In Minneapolis 64% of the eligible families who were originally asked to participate did so, and in Bombay the figure was 93%.

Experimental Task

The observed behavior data for this chapter are from an experimental study designed to involve families in an interesting and absorbing task. This task (known as the Simulated Family Activity Measurement, or SIMFAM) is a puzzle in a form of a game played with balls and pushers. The game is played on a court, about 9 by 12 feet, marked on the floor. There are two wood target boards at the front of the court (See Figure 1). Also, at the front of the room are a blackboard and three pairs of red and green lights mounted on a single board. One pair of these lights is for the husband, one pair for the wife, and one pair for the child.

The family plays this game for eight three-minute periods. After the family is seated, a white band is tied to the husband's wrist, a yellow band to the wife's wrist, and a blue band to the child's. The family is then told:

1. The puzzle to be solved is to figure out how to play the game.
2. If what you do is correct, a green light will be flashed.
3. If what you do is wrong, a red light will be flashed.
4. By noting which color light flashed, you can figure out the rules of the game and use this information to get as high a score as possible.

In addition to the lights, a family could judge how well it was doing from scores that were written on the blackboard after each three-minute trial. (A complete description of the SIMFAM experimental task and associated measures is given in Straus and Tallman, 1971).

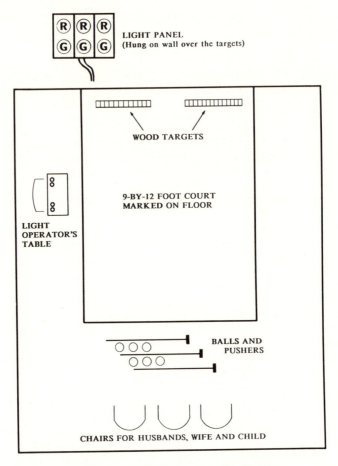

Figure 1. Floor Plan for SIMFAM

Measurement of Power

Simfam power score. Because there is almost no agreement on what is an appropriate method of measuring power within the family, three different techniques were used to measure power in each of the families studied. The first technique is based on observing the interaction of the family during the SIMFAM task. A tally was kept of each direction or suggestion made by the husband to the wife and whether she complied with that direction. The husband's SIMFAM power score consists of the number of such successful power assertions during the course of the experiment. A similar score was recorded for the wife.

Final say power index. The second power measure is a modification of the widely used technique developed by Blood & Wolfe (1960). Respondents indicted *"who had the final say"* for ten common family decisions using the following response categories:

> 1 = The wife only
> 2 = The wife more
> 3 = Husband and wife exactly the same
> 4 = Husband more
> 5 = The husband only

An index of marital power (which will be called the Final Say Power Index in this chapter) was obtained by summing the responses to the ten decisions. In the present research, Blood and Wolfe's technique was modified in three ways. First, since the respondents were children, the response categories were changed from "Wife" and "Husband" to "Mother" and "Father." Second, because all the families in this study had at least one child at home, two decisions concerning children were added to the eight Blood and Wolfe items. Third, since some of the original items were inappropriate for use in India (such as "What car to buy"), culturally appropriate family decisions were substituted.[2]

Immediate power index. The third technique is based on the pioneering work on measuring power by Herbst (1952). It consisted of a list of 12 day-to-day activities such as "Sees to it that I [the child] get properly dressed." The child indicated how often each of the 12 activities was the responsibility of each parent. This is followed by the same list of items asking who decides about that item, such as "Who decides what I should wear for the day?" The extent of each parent's decision-making activity is indicated using the following categories:

> 0 = Never
> 1 = Sometimes
> 2 = Half the time
> 3 = Usually
> 4 = Always

The sum of these 12 items for the father is called the Husband's Immediate Power Index in this chapter, and the equivalent measure is the Wife's Immediate Power Index.

Relative power. Preliminary analysis of this data indicated that the major difference in power structure between the Bombay and the Minneapolis families was in the SIMFAM power scores of the wife. Bombay husbands, as expected, did have higher power scores than did the husbands in Minneapolis. However, the largest difference was in the very low number of power acts emanating from the wives in Bombay. The most cross-culturally sensitive measure of marital power seemed to be the relative power of husband and wife. In Minneapolis, both spouses gave many directive acts to each other. The Bombay husbands seemed to have more power primarily because their wives gave few directions or suggestions during the course of the SIMFAM experiment.

To capture this aspect of marital power structure, a relative power score was computed for each couple by expressing the husband's power score as a percentage of the husband's plus the wife's power score. A similar procedure was followed for the Immediate Power Scores. Thus, a total of five power indexes were computed:

1. Husband's SIMFAM score
2. Husband's percent of the SIMFAM power scores
3. Husband's Immediate Power score
4. Husband's percent of the Immediate Power Scores
5. The Final Say Power score

Intercorrelations of Power Measures

The first concern to be raised about these five measures is the extent to which they measure the same thing. If they were highly correlated there would be no reason for using five different measures because one would suffice. But previous research on the relations between different measures of marital power indicates that the opposite problem is more likely; the different measures are essentially unrelated (Cromwell, Klein, & Wieting, 1975; Hadley & Jacob, 1973; Olson, 1969; Turk & Bell, 1972). Fortunately, as illustrated in Table 1, neither extreme prevails in this study. Although the size of the correlations vary, each of the measures is positively related to each of the others, but none of the correlations are so high as to indicate a redundant measure. Rather, it seems reasonable to assume that in addition to having something in common, each also taps a somewhat different aspect or facet of power (See Allen & Straus, 1984 for a more comprehensive analysis of measures of intra-family power). Consequently, for purposes of this chapter, a "Multidimensional Power Index" was constructed by standardizing each of the measures and summing the resulting scores.

Table 1. Intercorrelations of Five Conjugal Power Measures and a Multidimensional Power Index Combining the Five Measures

		IP-H	IP-%	FSP	SIMP	SIM%
MINNEAPOLIS						
IP-H	Immediate Power of Husband	–				
IP-%	Relative Immediate Power of Husb.	.46	–			
FSP	Final Say Power	.27	.61	–		
SIMP	SIMFAM Husband's Power	.43	.39	.14	–	
SIM%	SIMFAM Relative Power of Husb.	.30	.27	.15	.69	
MDPI	Multidimensional Power Index[a]	.48	.54	.31	.67	.56
BOMBAY						
IP-%	Relative Immediate Power of Husb.	.66	–			
FSP	Final Say Power	.08	.29	–		
SIMP	SIMFAM Husband's Power	.23	.19	.10	–	
SIM%	SIMFAM Relative Power of Husb.	.21	.35	.01	.34	
MDPI	Multidimensional Power Index[a]	.35	.53	.12	.28	.35

[a]Correlations with the MDPI have been corrected for inclusion of the item as part of the MDPI (see Nunnally, 1968:262).

Resources and Power

Table 2 lists the correlations of the Multidimensional Power Index with each of the resource variables available for this sample.[3]

Most empirical tests of the resource theory of marital power have been restricted to economic and occupational resources, as though these were the only things which husbands and wives had to offer each other. So we start with considering resources of this type even though there is considerable justification for Safilios-Rothschild's (1970) criticism of marital power studies because they omit non-economic resources.

Occupational and Educational Resources. The first row of Table 2 shows that in both Bombay and Minneapolis, husbands with white collar jobs have more power within the family than do husbands who are manual workers. This relationship is one of the most consistently found by research on family power and is, therefore, not a particularly unique contribution of the present study. However, the slightly higher correlation for the Bombay sample, which is contrary to the hypothesis, raised tantalizing issues that guided this study. It was assumed that even in a metropolis such as Bombay, the emphasis on kinship as a basis for status both within and outside the family would lessen the impact of occupational position as a resource.

Table 2.   Correlations of Resource Variable with
Multidimensional Index of Husband's Power

	r	
Resources	Bombay	Minn.
Occupation and Education		
White collar occupation of husband[c]	.31	.33
Occupational satisfaction of husband[c]	.25	.20
Occupational satisfaction, relative of husband	.15	-.07
Husband had second job	-.07	.06
Education of husband[c]	.26	.37
Education, relative of husband	.04	.17
Interpersonal Skills		
Support acts of husband (SIMFAM)[c]	.18	.23
Support, relative of husband (SIMFAM)[c]	.34	.24
Support by husband (Questionnaire)[c]	.31	.31
Esteem for father of child[c]	.09	.47
Esteem for father relative to mother[c]	.00	.40
Creativity of husband[c]	.51	.65
Creativity, relative of husband[c]	.39	.56
Task performance of husband[c]	.12	.49
Task Performance, relative of husband[c]	.19	.49
Age of husband[c]	.01	.28
Age, relative of husband	.09	.13
Family Structure		
Household type[bc]	.23	.22
Wife is not employed for wages[c]	a	.14
Wife's liking for being a housewife[c]	.21	.06

[a]No correlation was computed because there were only two Bombay wives in the paid labor force

[b]For Bombay this variable is nuclear household=1, Patrilineal or fraternal joint household=2, For Minneapolis, this variable is no person other than the nuclear family present=0, other persons present in household=1.

[c]Indicates varibles included in the multiple regression analysis (selected on the basis of at least one $r \geq .20$, i.e., .05 level of significance.

A correlation of .20 or greater was used as a cutting point since this is the smallest correlation that is significant at the .05 level with a sample of this size. Those resources for which a correlation of this size was obtained for at least one of the two societies are marked with a "c."

Two other variables under occupation and education, meet this criterion. They are husband's education and husband's occupational

satisfaction. For both these variables, similar co
obtained in Bombay and Minneapolis. Thus, each o.
resource variables which influence power in Minneap
to about the same extent in Bombay.

Interpersonal Skill Resources. Some of the variables under
section of Table 2 are based on observing the family interact during
the SIMFAM experimental task; others are based on questionnaire
data, and some, such as the level of interpersonal supportiveness of
the husband, are measured by both observation and questionnaire.
The SIMFAM observational measurement of this variable consists of
the number of times the husband praised, showed affection, consoled,
and helped other members of the family (See Straus & Tallman, 1971
for full details of the scoring method). The questionnaire measure
of support asked about these same types of behavior, and in
addition, included items dealing with companionship, i.e., spending
time with others in the family (See Straus, 1964a; 1974). The same
scores were computed for the wife, thus permitting a measure of
relative supportiveness of the husband (specifically, the husband's
percentage of the combined husband and wife scores).

But why consider interpersonal supportiveness as a resource on
which to base power? The reasoning behind this is simple--most of
us want to do things for people who are nice to us. If someone has
helped you or been supportive, then you are more likely to do something for them in return. This is roughly what French and Raven
(1959) call "referent power."

The first four rows under the heading, interpersonal skills, in
Table 2 show that a relationship exists between being a warm and
supportive husband and being able to influence others. This relationship holds for families in both Bombay and Minneapolis, regardless of which of the four different supportiveness indexes is used to
compute the correlation.

The fact that essentially the same results are obtained by using
two very different techniques (observing families interacting in a
laboratory and children's responses to questionnaire items about the
behavior of their parents) provides evidence of concurrent validity
(See also Allen & Straus, 1984).

The childs' esteem for the father (See Table 2, Interpersonal
Skills), is also a means of exercising "referent power," i.e., power
which is based on the person's identification with or attractiveness
to another person. It was measured by a question which asked the
child to tell how many things about the father he or she would like
to change. It was assumed that the fewer the changes desired, the
greater the child's esteem for the father. This index is among the
least adequate of the various resource measures for two reasons.
First, it is particularly sensitive to socially conventional responses.
Second, this chapter focuses on the balance of power between

etween husband and wife, but this balance is measure of the father's esteem through the child's eyes, not the wife's. Nevertheless, it does provide a measure of the influence of being a "referent" as a basis for exercising power.

The results in Table 2 indicate that the child's esteem for the father is strongly (and positively) related to the husband's power in Minneapolis, but no relationship is evident in the Bombay sample. Although the limitations of the data make interpretation hazardous, it is possible that because children in Bombay play a more passive role and exert relatively little influence in their families, their fathers might be less dependent on and sensitive to their actions and opinions (Straus, 1971).

The next resource shown in Table 2, creativity of the husband, can be taken as a resource for one aspect of what French and Raven (1959) call "expert power." Two previous studies investigated this variable. Centers, Raven and Rodrigues (1971) used "feelings of self-competence" and found no relationship with power. On the other hand, Smith (1970) used a measure of the child's belief in the parent's knowledge and found a strong relationship to the child's willingness to comply with the parent. As seen in Table 2, under interpersonal skills, there is a high correlation between the husband's creativity and the wife's level of compliance to his decisions and power assertions, regardless of whether creativity was measured by itself or relative to that of the wife (See Straus & Tallman, 1971 for the method used to measure creativity).[4]

The two indices measuring the "Family Task Performance" in Table 2 are based on the child's response to how often his/her father did each of the 12 family task items previously described. The positive correlations indicate that husbands who do things around the house and with the children tend to have greater power in relation to their wives. Evidently, for Minneapolis families, husbands who help with family chores are more likely to be listened to by their wives. The much lower correlations for the Bombay families suggest that, consistent with the "resources in cultural context" theory, household task performance is not an important basis for power for husbands. From an exchange theory perspective, the high correlation of household task performance with the husband's power can be interpreted as indicating that husbands who contribute heavily to household and child care tasks build up "credits" which the wife is motivated to repay by being receptive to such a husband's wishes. But there are other possible intervening processes.

For example, husbands who perform household tasks may be more competent than other husbands, and this competence may be the basis for both their ability to perform well and the influence they are able to exert on the behavior of their spouse (Kolb & Straus, 1974). Still another factor which might provide the link between family task performance and power is the operation of an implicit principle which gives to those who do specific things, the right to decide how they should be done.

These are all testable theories as to the process underlying the correlation between the husband's family task performance and his power. It will take additional research to find out which (if any) of these intervening processes is at work. Nonetheless, the husband's family task performance is related to his power in the family.

The final resource in the section on interpersonal skills (See Table 2) is the father's age. It was hypothesized that opposite trends would be observed in Bombay and Minneapolis for several reasons. First, the USA is widely held to be a youth-oriented and youth-admiring society which values creativity and curiousity in children and encourages them to express their opinions and knowledge. This is quite different from the parent-child interaction patterns in a society, such as India, where age connotes wisdom and children are expected to be submissive.

Second, while a wife's power relative to her husbands has been found to decrease when the family has young children (possibly because she is more dependent on him at this time) her power increases as the children reach adolescence and are available to form coalitions. Therefore, the older the father the more likely that all children in that family are older, thus increasing the wife's power relative to her husband's. On the other hand, Indian society is widely held to be one which venerates age. Consequently, the hypothesis underlying the use of husband's age as a resource variable predicted a positive correlation, i.e., the older the husband, the greater his power. Because of the wife and children's more submissive roles in India, we would expect them to exert little influence on this relationship. Therefore a negative correlation was expected for the Minneapolis sample between the husband's age and his power vis-a-vis his wife (Blood & Wolfe, 1960; Centers, Raven, & Rodrigues, 1971).[5]

The results are partly consistent with this hypothesis. For Minneapolis, both the absolute and the relative (to his wife) age of the husband resulted in the predicted negative correlations, i.e., the older the husband, the less his power. But in Bombay, both correlations were essentially zero. While this is not the predicted positive correlation, it at least demonstrates that, consistent with the presumed values of Indian culture, age is not the handicap to Bombay husbands that it is for Minneapolis husbands.

Family structural resources. Up to this point the influence of the larger social structure was mainly considered by comparing variables such as age and the effect on a husband's power in Bombay and Minneapolis. There is a negative effect of age on husband's power in Minneapolis and the absence of such an effect in Bombay. This illustrates the importance of considering "resources" in terms of their definition and meaning within a society. In contrast to this somewhat indirect method of examining "structural effects" on the husband's power in the family, the final section of Table 2 considers the effect of three family structural variables. The first of these concerns the household composition.

In Bombay, the hypothesis tested is that husbands in joint households have higher power (See Straus & Winkelmann, 1969; or Straus, 1974, 1975 for the classification procedures). This is because of the presumed greater attachment to traditional roles of those in joint households; the availability of adult relatives of the husband on whom he can depend for advice and help, and as allies in case of conflict with his wife; the corresponding relative isolation of the wife from such sources of interpersonal support; and the necessity for a clear line of authority in a large household. While these are cogent arguments, and there is some informal evidence consistent with this reasoning, an empirical test of these hypotheses is needed. The positive correlation for household type suggests, that even in the "modern" context of a city such as Bombay, the husband will have greater power if he is living in a household containing at least one other member of his lineage (See Table 2).

In Minneapolis, where the concept of the patrilineal joint household is absent as a cultural ideal, a related variable was examined: whether the household contains any adults who are not members of the nuclear family. Husbands living in families with other non-nuclear family members would be expected to have less power because such persons can restrict the husband's options and become allies of the wife in a dispute. The husband's inability to provide a house that is truly his private castle represents a failure to achieve the American cultural ideal of separate residence for each nuclear unit. Although the theoretical reasoning behind predicting a negative correlation for Minneapolis is not as clear as that underlying the expected positive correlation in Bombay, the results support the theory. Thus, in both cultural contexts, household composition exerts about the same effect on the husband's power, but as predicted by the resources in cultural context theory, in opposite directions.

Although it was not possible to compute a similar measure for the Bombay sample as only two wives were employed outside the home, in the Minneapolis sample, there was a slight tendency for the husband's power to be greater when the wife was not employed. Since this is a variable which has been shown in many studies to affect the husband's power, the small size of the correlation is surprising.

To further investigate this issue, the correlations were recomputed with social class controlled. Among the working class, husbands' power was highly correlated with wives' lack of employment (.36), but among the middle class the correlations were reversed (-.11). Wife's employment in middle class families did not affect the balance of power between the spouses. This finding is consistent with that reported for a different sample by Allen and Straus (1980), in which a lower correlation between resources and power was observed for the middle class than was observed for working class spouses.

Finally, the variable measuring the wife's enjoyment of being a housewife, which is structurally parallel to the wife's employment variable, was examined. The reasoning underlying the decision to correlate the extent to which the wife liked homemaking with the husband's power is that enjoyment of this role makes her more economically dependent on and committed to the marriage. In exchange theory terms, the alternatives are less attractive, or even psychologically unavailable.

Surprisingly, the expected positive correlation was found for the Bombay families but not for the Minneapolis families where the correlation was hypothesized to be stronger. The correlations with middle class families were -.01, but were .14 for the working class, further demonstrating the influence of social class.[6]

Combined Effects

The correlations discussed in the previous section follow the style of data analysis which has apparently characterized most previous research on the resource theory of marital power: correlating or cross tabulating resource variables with power one at a time. Using this mode of analysis, the findings of the present study are largely consistent with those of the previous studies; the relationships are consistent with the theory but tend to be quite low. Very little of the variation in the power of husbands is explained by any one variable.

Such a mode of analysis might lead one to an incorrect impression that little progress has been made in explaining why some families are relatively male-dominant and others are not. After all, a correlation of .25, even though statistically significant and replicated in several studies, accounts for only 5% of the variance in power. But such a perspective implies a single causal factor view of the determinants of social relationships. From a more realistic multiple causal view, a correlation of .25 is to be expected when a single independent variable is related to power. On the other hand, the same multiple causation perspective suggests that the combined effect of a whole set of resources might account for a considerable proportion of the variation in marital power. The technique used to examine the combined effect of the set of resource variables available for this sample is a multiple regression.[7]

The variables selected for inclusion in the regression analysis are those in Table 2 which have a statistically significant correlation with the husband's power in either the Bombay or the Minneapolis sample. The regression of these 17 variables on husband's power for Minneapolis produced an extremely high multiple correlation (.81), which means that approximately 66% of the variation in husband's power has been accounted for by this set of resource variables.

For the Bombay sample, the multiple correlation analysis was carried out twice. First, using the entire set of 17 variables, the multiple correlation was found to be .75. Although this is lower than the .81 obtained for Minneapolis, it is not as low as was expected on the basis of the resources in cultural context theory. However, using all 17 variables is not an appropriate test of the resources in cultural context theory because it includes the "nuclear household versus joint household" variable. To include this in the regression with the set of individual characteristics stacks the cards against finding support for the hypothesis that individual characteristics are a less important determinant of the husband's power in the context of Indian culture. Therefore, the more appropriate test requires that this variable be omitted from the regression. When this is done, the multiple correlation is .71 rather than .75. Consequently, the proportion of variance in husband's power explained by the remaining (essentially individual attribute) variables is 50%, which is somewhat lower than the 65% explained by the same regression equation for the Minneapolis sample. Nevertheless, the fact remains that 50% of the variation in husband's power is associated with the same set of variables as were found to influence the power of husbands in Minneapolis. Therefore, it seems reasonable to conclude that a similar set of social processes are at work in both cities despite the vast difference in the cultural context.[8]

SUMMARY AND CONCLUSIONS

This chapter presented data on the balance of power between husbands and wives in Bombay and Minneapolis. It tested an exchange theory of power by examining the extent to which various individual and structural "resources" of the husband influence his power in the family. The sample consisted of 64 husbands and wives in Bombay and 63 couples in Minneapolis. The data were obtained from two sources: observations of the couple's interaction with each other and one of their children while attempting to solve an engaging laboratory problem, and from questionnaires completed by the child some months earlier.

Because of the complexity and multidimensional nature of marital power, three different techniques were used to measure the power of husbands: A "final say power score" based on the methods developed by Blood and Wolfe; an immediate power score based on the methods developed by Herbst; the SIMFAM technique developed for this research. Since these three techniques produced low to moderate correlations with each other, they were combined to create a multidimensional power index.

This investigation was designed to overcome two of the deficiencies that previous research on the resource theory of power in families has encountered. First, there has been an almost exclusive preoccupation with economic resources. The resources in the present study also include interpersonal skills and family structural variables. Second, previous studies examined each resource in isolation from the others. By contrast, the present study relates all three groups of resource variables to the husband's power by means of multiple regression.

Of the 21 resource variables examined, 17 were found to be significantly related to the husband's power in either Bombay or Minneapolis. Some of the resources were reversed or more highly related in one society than in another, reflecting differences in the cultural definition and meaning of the same phenomenon in the two societies. For example, as expected on the basis of the youth orientation of American society, the older the husband the less his power; whereas, in Bombay, age did not have a negative effect on the husband's power (See Straus & Vasquez, 1975 for a another example of this type of reversal).

As is typical of most of the previous research, the correlations between any one of these independent variables and power tended to be low. However, when the combined effects of all 17 variables were estimated by means of multiple regression, multiple correlations of .63 to .81 were found for Minneapolis and .54 to .71 for the Bombay sample (See footnote 8). These multiple correlations, especially for Bombay, were larger than expected on the basis of Rodman's (1972) cultural context resource theory.

A lower correlation with power was expected for the Bombay families on the basis of the assumption that power in the Indian family system is primarily derived from one's position in the kinship network rather than from individual characteristics and achievements. However, recent research on families in kinship oriented patriarchal societies shows that even in this context, education is related to family structure (Lee, 1987; Miller, 1984). Moreover, as Sussman (1987) notes, the family is a highly adaptive and powerful institution. Families mold their environment and are responsive to the social environment. The findings of this study illustrate both processes. They show the effects of the cultural context on the way resources influence marital power, and suggest that, despite these cultural context factors, family structure changes in response to the realignment of the resources of husbands and wives in a similar way in all urban settings. ▩

Marvin Sussman's pioneering studies of the extended kin network and his cross-national comparative research were both important influences on my research. Therefore, it is a pleasure to be able to contribute this chapter to a volume recognizing Professor Sussman's contributions to family research. I wish to acknowledge Thomas Sparhawk and Patricia Craig for the data analysis, and Craig Allen, George Conklin, and Kersti Yllo for comments and suggestions on the first draft. This chapter is part of a three-society cross cultural study of families. A bibliography listing the books and papers is available upon request.

ENDNOTES

1. For purposes of this paper, power is defined as the control by one person of the behavior of another. Power, as used in this paper is, therefore, to be distinguished from such closely related phenomena as "power assertions," "power modality" (the specific type of act carried out as a power assertion), "power norms," and "power resources." See Rogers (1974) and Cromwell & Olson (1975) for a fuller discussion of power.

2. The items used in Minneapolis were: What car to get? Getting life insurance or not, and the value of the policy? Where to go on a vacation? What house or apartment to take? Whether mother should go to work or quit work? How much money the family can afford to spend for food? How the family income is spent in general? How the house is run (use of rooms, arrangement of furniture, interior decorating, etc.)? The parents' social and recreational activities (when to have company, whom to invite, what invitations to accept, whether and where to go for an evening, etc.)? Things concerning the children's activities (getting special privileges, discipline, staying out late, etc.)?

   The items used in Bombay were: Should a radio, furniture or a sewing machine be purchased? About getting insured or not, and the value of the policy? Where to go on a Sunday or any other holiday? Whether to go to a movie or not and which one to see? Should mother go to a job or should she leave it? What amounts should be spent on food grains, vegetables, and retail per month? On which things should the money from family income be spent? Things pertaining to the activities of the children (i.e., their privileges, staying late out of the home, discipline, studies, etc.)? Who should be visited? Who should be invited to a function or a festival? What should be given as a present? Should money be saved? If so, to what extent?

3. Each of the correlations to be reported in this paper was also computed using the five separate power indexes. Although the size of the correlations using the separate power indexes were generally lower (since each by itself is a less adequate measure of power), the central findings of the paper would not be changed had the separate power measures been used rather than the multidimensional power index.

4. The creativity score is the number of different ways of solving the experimental task offered by each member of the family (See Straus & Tallman, 1971 for the scoring manual). Since many of the possibilities will be described to other members of the family and also suggested to them, the creativity score overlaps with the power score. However, this is far from a complete overlap because many of the creative ideas were things that the person tried out without verbalizing, and others were verbalized but not in the form of a suggestion to others. Finally the power score used in this chapter is the number of directions and suggestions which the other person complied with, not just the number of such directive acts initiated. It should be noted, however, that part of the correlation shown in Table 2 might be due to this methodological artifact, as well as deference accorded to those recognized as creative.

5. This hypothesis may seem to be inconsistent with the theory that men in most societies marry women younger than themselves because it helps maintain a male dominant family structure since the husband is then "older and wiser." That principle, however, would seem to apply primarily at the time of marriage, where the difference between an 18-year old bride and a 22-year old husband can represent a significant difference in life experience and maturity.   But at the age of the spouses in this sample--all of whom had at least one junior high school age child--a difference in age of a few years becomes insignificant, or is overshadowed by the loss of youthfulness. Moreover, a dramatic reversal can occur late in the marriage when it is common for an aged husband to be physically dependent on his younger wife.

6. The potential contribution of family structural resources is probably much greater than is suggested by the limited number and type of such variables available for the present sample. For example, Conklin (1973) tested the theory that wives in cross-cousin marriages have greater power because the bride would arrive in a household in which she is not a complete stranger; because she is, after all, his cross-cousin; and because such marriages typically do not involve a long absence from her own parents' household. Conklin found the predicted difference in

wife's power. He also found that urbanization made an even greater difference, i.e., that urban wives had greater power than did rural wives.

7. Multiple correlation and regression assumes linear additive effects and it has already been shown that at least some variables have interactive effects. However, it did not seem advisable to add interaction terms to the equation because of the relatively small sizes of the two samples, and because of the multicollinearity problem which such terms usually create. Since interaction effects are not considered, the results presented may well underestimate the true combined effects.

8. Because of the possibility of measurement overlap discussed in footnote 4, the multiple correlations omitting the two creativity measure should also be considered. This results in multiple correlations of .60 for Bombay and .63 for Minneapolis. However, eliminating these two variables assumes that the entire correlation of creativity with power is due to measurement overlap. If, as was suggested in footnote 4, only part is due to this methodological problem, the "true" multiple Rs would be higher.

REFERENCES

Allen, C.M., & Straus, M.A. Resources, power, and husband-wife violence. In M.A. Straus & G.T. Hotaling (Eds.), *The social uses of husband-wife violence.* Minneapolis: University of Minnesota Press, 1980.

Allen, C.M., & Straus, M.A. 'Final say' measures of marital power: Theoretical critique and empirical findings from five studies in the United States and India. *Journal of Comparative Family Studies,* 1984, 15, 329-344.

Bahr, S.J., Bowerman, C.E., & Gecas, V. Adolescent perceptions of conjugal power. *Social Forces,* 1974, 52, 357-367.

Blood, R.O., & Wolfe, D.M. *Husbands and wives: The dynamics of married living.* Glencoe, IL: The Free Press, 1960.

Buehler, M H., Weigert, A.J., & Thomas, D. Correlates of conjugal power: A five culture analysis of adolescent perceptions. *Journal of Comparative Family Studies,* 1974, 5, 5-16.

Centers, R., Raven, B.H., & Rodrigues, A. Conjugal power structure: A re-examination. *American Sociological Review,* 1971, 36, 264-278.

Conklin, G.H. Urbanization, cross-cousin marriage, and power for women: A sample from Dharwar. *Contributions to Indian Sociology: New Series,* 1973, 7, 53-63.

Cromwell, R.E., & Olson D.H. (Eds.), *Power in families.* Beverly Hills, CA: Sage Publications, 1975.

Cromwell, R.E., Klein D.M., & Wieting, S.G. Family power: A multi-trait-multimethod analysis. In R.E. Cromwell, & D.H. Olson (Eds.), *Power in Families*. Beverly Hills, CA: Sage Publications, 1975.

French, J.R.P., & Raven, B. The bases of social power. In D. Cartwright (Ed.), *Studies in social power*. Ann Arbor: Research Center for Group Dynamics, Institute for Social Research, University of Michigan, 1959.

Hadley, T., & Jacob, T. Relationship among measures of family power. *Journal of Personality and Social Psychology*, 1973, 27, 6-12.

Heer, D.M. The measurement and bases of family power: An overview. *Marriage and Family Living*, 1963, 25, 133-139.

Herbst, P.G. The measurement of family relationships. *Human Relations*, 1952, 5, 3-35.

Kolb, T.M., & Straus, M.A. Marital power and marital happiness in relation to problem solving ability. *Journal of Marriage and the Family*, 1974, 36, 756-766.

Lee, G.R. *Family structure and interaction: A comparative analysis*. NY: Lippincott, 1982.

Lee, G.R. Comparative perspectives. In M.B. Sussman & S.K. Steinmetz (Eds.), *Handbook of marriage and the family*. NY: Plenum Press, 1987.

Miller, K. The effects of industrialization on men's attitudes toward the extended family and women's rights: A cross-national study. *Journal of Marriage and the Family*, 1984, 46, 153-180.

Nunnally, J.C., Jr. *Psychometric Theory*. NY: McGraw-Hill, 1968.

Olson, D.H. The measurement of family power by self-report and behavioral methods. *Journal of Marriage and the Family*, 1969, 3, 545-550.

Olson, D.H., & Cromwell, R.E. Power in families. In R.E. Cromwell & D.H. Olson (Eds.), *Power in families*. Beverly Hills, CA: Sage Publications, 1975.

Price-Bonham, S. A comparison of weighted and unweighted decision-making scores. *Journal of Marriage and the Family*, 1976, 38, 629-640.

Rodman, H. Marital power and the theory of resources in cultural context. *Journal of Comparative Family Studies*, 1972, 3, 50-70.

Rogers, M.F. Instrumental and infra-resources: The bases of power. *American Journal of Sociology 79*, 1974, 1418-1433.

Safilios-Rothschild, C. The study of family power structure: A review-1960-1969. *Journal of Marriage and the Family*, 1970, 32, 539-552.

Smith, T.E. Foundations of parental influence upon adolescents: An application of social power theory. *American Sociological Review*, 1970, 35, 862-873.

Straus, M.A. Measuring families. In H.T. Christensen (Ed.), *Handbook of marriage and the family*. Chicago: Rand McNally, 1964.(a)

Straus, M.A.   Power and support structure of the family in relation
   to socialization.   *Journal of Marriage and the Family*, 1964, 26,
   318-326. (b)

Straus, M.A.   Phenomenal identity and conceptual equivalence of
   measurement in cross-national research.   *Journal of Marriage and
   the Family*, 1969, 31, 233-239.

Straus, M.A.   Methodology of a laboratory experimental study of
   families in three societies.   In R. Hill & R. Konig (Eds.), *Families
   east and west*. Paris: Mouton, 1970.

Straus, M.A.   Social class and sex differences in socialization for
   problem solving in Bombay and Minneapolis.   In J. Aldous *et al.*,
   (Eds.), *Family problem solving*. Hinsdale, IL: Dryden Press, 1971.

Straus, M.A.   Some social class differences in family patterns in
   Bombay.   In G. Kurian (Ed.), *The family in India: A modern re-
   ional view*. The Hague: Mouton, 1974.

Straus, M.A.   Husband-wife interaction in middle and working class
   nuclear and joint households in Bombay. *Studies in Honor of K.M.
   Kapadia*. Bombay, India: Thacker and Co., 1975.

Straus, M.A., & Tallman I.   SIMFAM: A technique for observational
   measurement and experimental study of families.   In J. Aldous *et
   al.*, (Eds.), *Family problem solving*.   Hinsdale, IL: Dryden, 1971.

Straus, M.A., & Vasquez M.M.   Adolescent-parent conflict in Bombay
   and Minneapolis in relation to social class and family roles.
   Sociologia Internationalis, 1975, 13, 171-190.

Straus, M.A., & Winkelmann, D.   Social class, fertility, and authority
   in nuclear and joint households in Bombay.   *Journal of Asian and
   African Studies*, 1969, 9, 61-74.

Sussman, M.B.   From the catbird seat: Observations on marriage and
   the family.   In M.B. Sussman & S.K. Steinmetz (Eds.), *Handbook of
   marriage and the family*. NY: Plenum Press, 1987.

Sussman, M.B., & Cogswell, B.E.   Interpersonal competence: An issue
   in cross-national family research.   In M.B. Sussman & B.E. Cogswell (Eds.), *Cross-national family research*.   Leiden, Netherlands:
   E.J. Brill, 1972.

Szinovacz, M.E.   Family power.   In M.B. Sussman & S.K. Steinmetz
   (Eds.), *Handbook of marriage and the family*. NY: Plenum Press,
   1987.

Turk, J.L., & Bell, N.W.   Measuring power in families.   *Journal of
   Marriage and the Family*, 1972, 34, 215-222.

▓ 12 ▓

INTERGENERATIONAL TRANSACTIONS OF VALUES: A CASE STUDY OF CAREER CHANGES IN MIDLIFE

Gail G. Whitchurch

INTRODUCTION

Since the late 1960s scholars have recognized that values flow not only from parent to child, but from child to parent as well (Bell, 1968; Rheingold, 1969). Values refer to an individual's subjective evaluations about the relative degree of desirability or undesirability of a particular variable, and can be influenced through relationships with other people, especially family members. According to Burr, Leigh, Day & Constantine (1979:49), *"values are assigned in the course of interaction and are constantly changing and tentative."*

THE TRANSACTIVE PERSPECTIVE

The transactive perspective developed in reaction to limitations of linear models, which perceive an individual's action as causing a re-action in another person, who then gives a response--much in the way a machine goes through a series of operations that bring about the desired output. The actions and re-actions are said to be *linear* because they extend along a line, moving in one direction at a time. A linear model of the flow of values among generations would focus on the similarities of values among generations, as in values flowing "up" to an older generation or "down" to a younger generation.

A transactive view takes into account the participants' *simultaneous* contributions to the process by which values flow among generations. In this perspective, family members are simultaneously perceiving each others' values and are adjusting their own values.

155

The *transactive* aspect lies in observing the full system without attributing cause to any single element (Dewey & Bentley, 1949). Transactions among individual family members form the family system, which is characterized as *open* or *closed*. The relative amount of openness the family permits in the system will regulate the boundaries around the individual family members and around the family system. In a closed system, a family would maintain rigid boundaries around the family unit, emphasizing similarity in values among generations. Individuals would be discouraged from developing idiosyncratic value structures. An open system--a family with permeable boundaries--would permit, or even encourage, members to discuss their values with individuals outside the family system. It would allow questioning of values held by individual family members.

Changes in the system would be determined by whether the family is an open or closed system. Any deviation from a closed system's steady state would be met by negative feedback making the system regain its steady state. Thus, a closed family system would resist the effects of events in historical time that bring about rapid changes in social values, such as those precipitated by a war, a depression, sudden prosperity, etc. Such a family might even reinterpret its own family history to fit its existing value structure. Family members might be enjoined from mentioning a premarital pregnancy, for example. The system would generate this type of negative feedback whenever the family members perceive that the system is out of balance, forcing the system to regain its steady state.

An open system would allow, or even encourage, its members to examine their own values, both inside and outside the family system, such as children being permitted to have friends with values different than the parent(s). This positive feedback would increase the divergence of the system from its former steady state, eventually recalibrating itself to a new steady state.

CHALLENGES TO THE TRANSACTIVE APPROACH

A transactive approach is not without its critics. For example, in discussing four ways that reciprocal interaction may be studied, Klein, Jorgensen and Miller (1978) discuss a transactive approach, which they term a "synchronous" approach:

> This synchronous view is also intuitively appealing, but it must cope with the problem of indeterminacy. Since it is infeasible to distinguish cause from effect when interpersonal influences are treated as simultaneous, someone who adopts this model would probably deny that causality is important or would claim that it cannot be empirically untangled. Methodologically, a synchronous perspective

*implies that family relationships can only be studied
using qualitative and descriptive methods* (pp.108-109).

Is it necessary, or even desirable, to "distinguish cause from
effect" for *all* conceptualizations of intergenerational transmission of
values? It may well be that two parallel research efforts on the
flow of values among generations must develop. One would be ongoing quantitative research, with its advantages of larger sample
size and suitability for statistical analysis. The other would be
qualitative and descriptive, as Klein *et al.*, (1978) have suggested.
This would have the advantages of tapping issues of context, as well
as transactiveness in the flow of values. The case presented below
illustrates the application of intergenerational transactions of values
to gain insight into the interactions between family members.

MARGARET[1]

Background

Margaret, a 48-year-old white woman, married 29 years to her
56-year-old husband, was the subject of this case study. Intensive
interviews were conducted over a three-month period, including an
all-day home visit by the researcher, as well as weekly face-to-face
and telephone interviews ranging from a half-hour to an hour each.
Margaret and John first met when she was 9 and he was 17,
and married 10 years later. Both were ready to fall in love,
according to Margaret, with *"storybook fantasies of love and
marriage."* They moved to their present rural community right after
their marriage. John owns an auto repair shop, and although they
are working class, Margaret noted that her neighbors think that
they are rich because of the high level of unemployment and poverty
in their neighborhood.
Margaret had begun attending college classes for the first time
on a part-time basis soon after her children left home for college or
a job. She eventually enrolled full-time in a degree program which
she will complete in one more semester. Margaret described herself
as having been a traditional, stay-at-home housewife and mother
before attending college, but has held a contract as a school bus
driver for the last sixteen years. She continued to drive the bus
during the time of the interviews, although she subcontracted most
of the work to her daughter, Andrea.
Marsha, 28, is the oldest child. She is employed as a chemical
sales representative, and she and her husband do not plan on
having children. Andrea, 27, is married and has an infant son,
Scott. She did not go to college and has held a number of non-
traditional jobs such as construction work. John Charles, 25, is a
master's degree program at another university.

Margaret grew up in a rural area of a nearby state. Her father died when she was eleven, so she and her several sisters were brought up by her mother, with the help of Margaret's two unmarried aunts. Margaret's mother supported the family by working in real estate, having unexpectedly earned her first license in the 1940s. She had helped Margaret's father study for the license examination and at the last minute decided to take the test. She passed the exam, but Margaret's father had to take the test a second time to pass.

Both parents sold real estate in the 1940s and 1950s. Margaret's father eventually passed the examination to obtain a broker's license, but Margaret's mother failed that exam *deliberately*, according to Margaret. Although her mother was a successful businesswoman, Margaret believed her mother was, unknowingly, the victim of sex discrimination. Margaret's mother is currently in a nursing home and Margaret writes three times a week and visits occasionally. In spite of the additional activities related to her new roles as college student and grandmother, Margaret works at keeping her close ties with her extended family, providing further refutation of the "isolated nuclear family" (Sussman, 1959).

Historical time. Margaret reported being so preoccupied with raising three young children that she was largely unaware of the political and social turmoil of the 1960s. Even so, Margaret was influenced in the 1980s by the historical legacy of the 1960s-- particularly the modern women's movement. Margaret entered midlife in a period of history in which it was not only acceptable, but even common, for women in midlife to attend college. A generation earlier, it is unlikely that this option would ever have occurred to her, or been feasible.

Cohort effects. Hagestad (1984) points out the importance of each family's combinations of age and cohort memberships, which reflect timing of births in several generations. Since Margaret began having her children in her late teens and ended in her early twenties, her children became young adults when she was in her late thirties. Even though Andrea waited until she was twenty-six to conceive, Margaret was still a much younger grandmother than if she had borne her own children in her thirties, and they had given birth in *their* thirties.

As a young grandmother, she found herself torn between the conflicting demands of two roles. Her cohort experience as a young mother in the late 1950s is evident in her feelings of schism between her student role and her cohort's idea of what a woman "should" do: the traditional role as stay-at-home mother, and then "full-time grandmother," who would cook meals for Andrea, Peter and Scott, and stay with Scott when needed. During the 3 months that Margaret participated in this study, she occasionally expressed

her concern that perhaps she should quit school and drive the
school bus so that Andrea could stay at home full-time with her
baby.   She eventually discarded that idea because she believed
"Andrea would feel ripped off" if she did that.   Thus, Margaret's
cohort experience was mediated by women holding paid jobs in the
historical time of the 1980s: *"Let* [Andrea] *have her life,"* said
Margaret.

Family Life Cycle

Margaret and John are still near the beginning of the "middle-
aged parents" or "empty nest" stage of the family life cycle (Duvall,
1977:148).   Because they had their children when they were young,
they can expect to live a large number of years in this stage.
Although John plans to continue to do auto repair indefinitely,
Margaret was faced with midlife decisions while still in her early
forties.   According to Margaret, the usual scenario in their
community would be for her to do church work with the time she
once devoted to her children.   There is fierce resistance among her
peers to her departing from this mold: "You're a country girl and
you always will be, and don't you ever forget it," a neighbor told
Margaret.   However, Margaret's neighbor didn't know about Lillian
Rubin, who, Margaret explained, was a major influence:

> *While I was between the ages of 40 and 45, my best
> friend had to be Lillian Rubin.   She explained to me in
> her book,* <u>Women of a Certain Age</u>, *the disparities of the
> men's and women's roles in the 'total relationship' so
> highly praised in love stories.   She was my friend while
> she told me of the need for many friends on different
> levels of networking.   Lastly, she told me about. . . what
> midlife is for 'family women.'   My friend, Lillian, told me
> that the empty nest need not be a time for endings; it
> could be a time of new beginnings.*

Entering the "middle-aged parents" stage of the family life
cycle allowed Margaret the option of other activities for her
remaining years, but her statement above clearly indicates that life
stage alone does not account for the choices she made.   Her
community peers and her social class constrained her traditional
choices.   As Rubin (1979) points out, working class girls are likely
to have limits set on their educations soon after they enter high
school.   Margaret reports her mother was anxious to "marry her off"
as soon as possible after high school because she feared Margaret
would get pregnant.   Even so, Margaret eventually secured a college
education not only for her children, but for herself.   The apparent
contradiction could be accounted for by her family's unique history.

Unique Family History

Even though Margaret might have otherwise been constrained from a college education, and locked into the traditional "full-time grandmother role" by her working class roots and her peers, she combined a college career with a modified grandmother role. The roots were there in the unique history of Margaret's family of origin: Margaret's mother, who "at the last minute" took the real estate examination and worked for pay even before her husband's death, unusual in the years before World War II.

The unique history of Margaret's family of procreation also provides ample evidence that social class alone does not account for values. One incident stands out. When Margaret and John had moved from the state in which they were married to their current home, their eldest daughter, Marsha, had been placed in what was known as the "dumb" class at the elementary school, simply because she was new. Margaret realized this would peg Marsha forever, so she sat on the school steps until the school principal agreed to give Marsha a chance in the "smart" class. From then on, Marsha had a reputation for being a bright kid and was always placed in the "smart" class.

Even though neither Margaret nor her parents had had a college education, Margaret recognized the impact that having her daughter labeled as "dumb" would have on her future opportunities to attend college. Margaret reported that because Marsha thought of herself as a "smart kid," she always assumed she would attend college, and eventually the behavior that stemmed from that value--attending the university--later passed back to Margaret when she attended the university herself. This is in line with findings of substantial parent-child agreement in values regarding interest, plans and motivation (Bengtson & Troll, 1978).

The school placement incident is a microcosm of the unique history of Margaret and John's family of origin. Margaret tells how much she enjoyed talking about current events and philosophy with her teenage children. That led to her decision to attend the university after all the children had left home:

> Marsha, and then John Charles, had gone to the state university, so for me there was no school but the state university. . . soon, I was getting more out of school than out of driving the bus!

The family's unique history is also evident in the values that led to Andrea's high priority on motherhood. Margaret and Andrea spent many hours together after Andrea's marriage--giving advice to one another, sharing their ideas and hopes during a walk every evening--more like women friends than mother-daughter, according to Margaret. Mother-daughter interdependence is evident in Margaret's involvement in Andrea's pregnancy, even before conception:

> *In November [1986] we did a home pregnancy test.*
> *Negative. Oh boy! We were disappointed. . . For the*
> *first time in my life, I began to really feel what it must*
> *be like for a married couple to be unable to conceive.*
> *Baby was all Andrea and I talked about. Did I remember*
> *how I felt? Was I cranky? Did I feel sick mornings?*
> *How much, how soon, how often?*

Then, when the home pregnancy Margaret bought for Andrea was positive:

> *Well, the test showed positive! Oh, the excitement.*
> *Peter and I danced and hopped and yelled around their*
> *kitchen. We thought up names, made plans, had the baby*
> *born, named, educated and married--all in a Saturday*
> *morning.*

Individual Factors

Individual development. Individual stages of development can be characterized in a number of ways. Some of Margaret's developmental needs are being met by grandmotherhood:

> *Now that I knew I could be a grandmother--I was so*
> *happy! I didn't want my daughters to want a baby just*
> *so I could be a grandmother but now I could be a*
> *grandmom! Good!*

Margaret has midlife needs in addition to grandmotherhood. Her need to be challenged by instrumental activities during her middle years (Rubin, 1979) is being met by work toward her bachelor's degree. Each course is a goal in itself as Margaret works toward the ultimate goals of a obtaining her degree and paid employment in a career based on that degree.

One's own value system. During the three-month period of the interviews, Margaret's own value system was modified to include her new role as grandmother. The process of re-ranking her roles was evident during the interviews. For example, when she had promised to go with Andrea to have Scott's first photograph taken before she realized she had a paper due the next day, Margaret chided herself: "*Get your priorities straight.*" When asked what her top priority was, she replied, "*Scott!*" without hesitation.

However, by the end of the three months of interviews, Margaret had settled on dual priority on family roles and responsibilities, as well as on her college degree. This was evident some two months after the incident with Scott's photograph, when she described herself as happier since she entered college than before she enrolled. She spoke of women who attend college in midlife: *"We have to do this."*

System change. The extended family system recalibrated with Margaret's changes, especially her college work: her married daughters brought cooked dinners to stock Margaret and John's refrigerator during final exams. John's value system in particular seemed to be affected by changes in the family system. He began to feel a responsibility towards assuming the home chores he had refused to do before Margaret entered the university.

Margaret's report of her own increased happiness is not surprising in light of research findings that working wives' morale is highest in families where husbands share in meal preparation and household tasks (Clausen, 1986). John's individual needs may have been met in the process. At 56, he was closer to retirement than Margaret and was looking forward to the security that her post-college benefits would provide.

CONCLUSION

The boundaries both within and around Margaret and John's family are clearly flexible, those of an open system. When Margaret began to think of attending college, she discussed her ideas with her children first. This is not unexpected, given the findings of Hagestad (1977), that children often serve as bridges to nonfamily spheres such as work or education, and of Hagestad (1984) that in later adulthood, women frequently use adult children as confidants.

It is impossible to identify whether Margaret's children took on her values on education, or if she took on theirs, as would be the goal of a linear model. When seen through a transactive model, however, new insights are gained as the model reveals the interconnections not only in the individuals, but in their value systems, within the broader context of historical time, family life cycle and the family's unique history. ▨

ENDNOTES

1. In order to maintain the anonymity of this family all names and identifying information have been altered.

REFERENCES

Bell, R.Z.   A reinterpretation of the direction of effects in studies of socialization. *Psychological Review*, 1968, 75, 81-95.

Bengtson, V., & Troll, L.   Youth and their parents: Feedback and intergenerational influence in socialization.   In M.L. Richard & G.B. Spanier (Eds.), *Child influences on marital and family interaction: A life-span perspective.* New York: Academic Press, 1978.

Burr, W.R., Leigh, G.K., Day, R.D., & Constantine, J.   Symbolic interaction and the family.   In W.R. Burr, R. Hill, F.I. Nye & I.L. Reiss (Eds.), *Contemporary theories about the family.* (Vol. 2) New York: Free Press, 1979.

Clausen, J.A.   *The life course: A sociological perspective.* Englewood Cliffs, NJ: Prentice-Hall, 1986.

Dewey, J., & Bentley, A.F.   *Knowing and the known.* Boston: Beacon Hill Press, 1949.

Duvall, E.M.   *Marriage and family development* (6th ed.). Philadelphia: J.B. Lippincott, 1977.

Hagestad, G.O.   Role change in adulthood: The transition to the empty nest. Unpublished manuscript. Committee on Human Development. University of Chicago, 1977. (Cited by Bengtson & Troll, 1978)

Hagestad, G.O.   The continuous bond: A dynamic intergenerational perspective on parent-child relations between adults.   In M. Perlmuter (Ed.), *Parent-child interaction and parent-child relations in child development.*   Minnesota Symposium on Child Psychology, (vol. 17). New York: Erlbaum, 1984.

Klein, D.M., Jorgensen, S.R., & Miller, B.C.   Research methods and developmental reciprocity in families.   In R.M. Lerner & G.B. Spanier (Eds.), *Child influences on marital and family interaction: A life-span perspective.* New York: Academic Press, 1978.

Rheingold, H. L.   The social and socializing infant.   In D.A. Goslin (Ed.), *Handbook of socialization theory and research.*   Chicago: Rand McNally, 1969.

Rubin, L. B.   *Women of a certain age: The midlife search for self.* New York: Harper Colophon, 1979.

Sussman, M.B. The isolated nuclear family: Fact or fiction.   *Social Problems*, 1959, 5, 333-340.

▩ 13 ▩

PARENTAL AND FILIAL RELATIONSHIPS: OBLIGATION, SUPPORT AND ABUSE

Suzanne K. Steinmetz

INTRODUCTION

There are proverbs available for any of life's circumstances. For example, the proverb *"The grass looks greener on the other side,"* warns us that whatever our situation, it is inevitable that another will seem much more appealing. Thus, as individuals grapple with the stresses and conflicts of daily life, it is understandable that another time, another set of circumstances, another's life will certainly appear to be more enticing.

During the last century, we have seen formal institutions increasingly replace the family as providers of religious, educational and occupational training. The principal responsibility of the family remains the fulfillment of the expressive needs of its members. Therefore, the way in which care is provided, not merely the provision of care, has become the yardstick by which we measure the health of family life. In this chapter we will explore both the myth and reality of parent-child obligations and their fulfillment over the life cycle and over time. First we will examine the myth and reality of family life from a historical and contemporary perspective. Then we will focus more specifically on the treatment of children and elders over time.

The Family of Yesteryear

In our attempt to glamorize the family of the past we have overlooked its failures and idealized its strengths. We have tended to blame contemporary family related problems on changes such as increased occupational and geographic mobility, rising divorce rates,

165

the two car and two career family, and the "pill." But there is abounding evidence to refute the notion of idyllic intergenerational relationships in families of the past.

Hareven (1977:65) observes that: *"Families shared their household space with other kin only as a last resort during periods of housing shortages or severe economic constraint."* The family idealized in literature was not the reality experienced by most families. Kent (1965:55) notes, for example, that:

> *The three-generation family pictured as a farm idyll is common, yet all evidence indicates that at no time in any society was a three generation family ever the common mode, and even less evidence that it was idyllic.*

Further articulation of the above can be found in Fischer's (1977) analysis of demographic life cycle data from 1650 to 1950. Until 1850, women, whose life expectancy was about 50, would have been in their late 50s or early 60s before their last child came of age. However today, with a life expectancy for women close to 80, and women concluding their childrearing responsibilities while in their early to mid 50s, women could expect to experience approximately 33 years without the responsibility of traditional roles affiliated with motherhood. Women in Pre-Industrialized America did not have to face the "empty nest" stage. Unlike the myth of the three-generation family of yore, these women (and their husbands) spent the majority of their old age in their own nuclear family, not in the family of their adult child. The above statistics can best be illustrated by Haber's (1983:10-11) analysis of the Thomas Willing family over three generations:

> *Thomas Willing, a Philadelphia merchant and politician was born in 1731, the oldest of eleven children. His youngest sister, Margaret, was born in 1753, 23 years and an entire generation later . . . From 1764 (a year after his marriage) to 1781 (the year of his wife's death), Willing and his wife had thirteen children . . . In 1781, the year their daughter Anne married William Bingham, Thomas Willing and his wife had their final child . . . In 1799, Anne and William Bingham became grandparents for the first time; a year later, their own son was born.*

Thomas Willing's parents would have spent 40 years raising a family, had they survived. Thomas approached his seventh decade before his last surviving child reached adulthood, and the birth of his own daughter's first and last child was separated by nearly two decades. Parental roles and responsibilities remained until advanced age and, in fact, were often not completed at the time of death.

In addition to the above, historical evidence documents the

general mobility patterns in which successive generations moved westward to the new frontiers. The young moved to the new areas and the elders remained at home, whether it was in the "old country" or the recently developed areas of the U.S.

Therefore, if a three-generational pattern did exist in a given family, it tended to be for a short span of time--while a young couple built their home, or amassed resources needed for settling new lands, or when the elder moved in to die. Furthermore, the shorter life expectancy decreased the overlap between generations. Only the fittest survived; the weak and infirm, young and old without recourse to modern medicine succumbed quickly. Contrary to the picture of the family gathered at the bedsides of the elder devoting 24 hours a day to their care, the elderly continued working until severe illness or death, and were not dependent on their adult children for prolonged care.

Today a sizeable number of persons over 65 have one or more living parents; one in ten elderly have a child over sixty-five. Brody (1978) reports that approximately 82% of the elderly have at least one surviving child; 70% have grandchildren; 30% have great-grandchildren. This trend, however, is not without drawbacks. He warns (1987:16) that:

> . . . in the future, not only will those who are dependent be older and therefore will require more services than older people in general, but those on whom they depend . . .their family members . . . also will be older . . . more often those depended on will be middle-aged, aging and even old themselves, with their capacities to be supportive diminished.

The romantic myth of the warm, loving, multigenerational family differs considerably from reality. In fact as Sussman (1959) eloquently argued nearly three decades ago, the contemporary nuclear family, although residentially separate, remains the primary unit of support of the older generation and maintains active, reciprocal support over the life cycle. Furthermore, contemporary treatment of the elderly, both by society and the family, is considerably more caring and humane than in earlier eras (Steinmetz, 1988).

Historical Treatment of Children

In Roman times, the Patria Potestae permitted fathers to sell, sacrifice, mutilate or kill offspring (Radbill, 1968). The practice of burying children in the foundations of buildings existed throughout history, and as recently as the 16th century German children were buried alive beneath the steps of public buildings (Bates, 1977). During the Industrial Revolution, children endured long hours, extreme physical hardship, and numerous beatings to ensure that they would exhibit no laziness. A historical analysis of the treatment of children (deMause, 1974), documents the cruel, almost non-human treatment of children in the past.

Colonial America often turned to the Bible for family guidance. Parents were advised that "*He who spares the rod hates his son, but he who loves him disciplines him diligently*" (Proverbs 13:24). Unfortunately, religious guidance on parental and societal control of children is unbelievably harsh. In fact, today much of this treatment would be deemed cruel and inhuman treatment and clearly a case of child abuse. For example, children were considered to be the property of their parents in the Hammurabi code of 2100 B.C. and the Hebrew code of 800 B.C. As property, parents had the right to decide whether they lived or died, and infanticide was considered to be an acceptable practice (Bates, 1977).

Child abuse, although only recently "discovered" by researchers, has its roots in the early history of our country. A 1646 law, in an attempt to help parents control their rebellious children, decreed:

If any child(ren) above sixteen years old and of sufficient understanding shall curse or smite their natural father or mother, they shall be put to death, unless it can be sufficiently testified that the parents have been very unchristianly negligent in the education of such children, or so provoked them by extreme and cruel correction that they have been forced thereunto to preserve themselves from death or maiming. . . If a man have a stubborn or rebellious son of sufficient years of understanding, viz. sixteen, which will not obey the voice of his father or the voice of his mother, and that when they have chastened him will not harken unto them, then shall his father and mother, being his natural parents, lay hold on him and bring him to the magistrates assembled in Court, and testify to them by sufficient evidence that this their son is stubborn and rebellious, and will not obey their voice and chastisement, but live in sundry notorious crimes. Such a son shall be put to death (Bremner, 1970:37).

It is of interest that our concern for protecting animals resulted in legislation that was later used to protect a child against abuse by her parents. In 1874, nine year old Mary Ellen was judged to be a member of the animal kingdom so that she could be protected from her abusive parents under the recently enacted laws protecting animals. Public reaction to the need to use the Society for the Prevention of Cruelty to Animals to protect children was instrumental in the later founding of the Society for the Prevention of Cruelty to Children. However, nearly a century passed before all states had laws mandating the reporting of child abuse, requiring investigation of the reported cases, and providing services to these children and their families.

Historical Treatment of Elders

The Bible also served as a guidepost for children's treatment of their parents. Children were commanded to "*Honor thy father and thy mother that their days be long upon the earth*" (Exodus 20:12). However, the respect enjoined by the fourth commandment did not necessarily embrace affection among colonial Americans. Although the elderly might have special pews in church and even enjoy a presumption of grace because of their wisdom, many were viewed as wicked, possessed of the spirit of the devil, or addicted to witchcraft. The society's remedies for unacceptable or suspect behavior were usually harsh. The extreme cases were resolved in the legendary witchcraft trials, many of which resulted in a sentence of death.

The loss of power and prestige with advancing age has been lamented throughout time. Increase Mather complained of loss of esteem after leaving the pulpit; his status of veneration and adoration now categorized as useless and outdated. In his essay, *Two Discourses* (1716:120), he noted "*it is a very undesirable thing for a man to outlive his work.*" His son, Cotton Mather, publicly berated the elderly for refusing to voluntarily retire from important positions, and in his essay, *A Brief Essay on the Glory of Aged Piety* (1726:28), he wrote:

> *Old folks, often can't endure to be judged less able than ever they were for <u>public appearances</u>, or to be put out of offices. But good, sir, be so wise as to <u>disappear</u> of your own accord, as soon and as far as you lawfully may. Be glad of a <u>dismission</u> from any post, that would have called for your activities.*

Not only were the elderly viewed as an undesirable blight on society, there is considerable evidence that they were neglected and abused. Cotton Mather, in *Dignity and Duty*, complained that "*There were children who were apt to despise an Aged Mother.*"

More than a century later, in 1771, Landon Carter recorded in his diary:

> *It is a pity that old age which everybody covets and everybody who lives must come to should be so contemptible in the eyes of the world* (Cited in Smith, 1980:275).

Haber (1983) has suggested that the elderly were classified according to very specific criteria that distinguished useful old age from superannuation. Those who were able to retain their authority and maintain the respect of the community were considered to represent *useful* old age. Those who lost status, power and wealth, the superannuated, were ridiculed. In her review of mid-19th century reform movements and charitable organizations, Haber asserts that the ideologically justified refusal of these organizations to assist the old were based on:

▓ Intractability of the character of the old (they would never change and therefore would never be worthy of charity); and

▓ They had outlived their years of maximum productivity, thus charity would not restore them to productive, worthwhile lives.

The elderly who were poor, female, and widowed, three attributes often found together, were despised and treated badly. Colonial court records disclose frequent attempts to bar old people from a town lest they increase the number of paupers. A New Jersey law passed in 1772 required the Justices of the Peace to search arriving ships for old persons as well as other undesirables and to send them away. Neighbors often "warned-out" poor widows and forced them to wander from town to town.

The minutes of a meeting of the Boston City Selectman held in 1737 reported:

Whereas One Nicholas Buddy an Idle and Poor Man has resided in this Town for Several Years past and is in danger of becoming a Charge of the Town in a Short time, if not Transported. And There being now an Offer made by some of his friends of Sending him to Jersey (his Native Country) Provided they might be Allowed the Sume of five Pounds towards defraying the Charges of his Passage tither (Records of the Boston Selectmen, 1936).

The Selectmen appointed Captain Armitage and Mr. Clarke to complete the arrangements and authorized payment, not to exceed five pounds, on condition that Nicholas Buddy be sent home.

We need to ask: *"Where were the loving, caring children of yore and why were they not caring for their elders?"* First, different standards were used to define appropriate treatment in colonial America. Those citizens enjoying the highest standard of living at that time would be considered to be living in abject, unacceptable, poverty by today's standards. When an elder was forced to live in an out-building, or wander about the town, this treatment was barely different from that experienced by many citizens representing all age groups. Second, it is quite probable that these elder had no children who were able to provide care, a situation that parallels contemporary society in which more than half of the residents of nursing homes have outlived family and friends (Nursing Home Care in the U.S., 1982).

However, there is evidence to suggest that, contrary to our view of elder-adult child bonds being close, loving, and supportive in colonial America, parents often resorted to legal means to protect their property and ensure that they were cared for in their declining years. For example, parents often found it necessary to use property transfers, both by will and by deed of gift as hedges against maltreatment of themselves or their survivors in their old age. These documents often included elaborate instructions for the surviving wife's care, and required the child who inherited the property to provide food, clothing, shelter, and services or risk forfeiture of the inheritance.

In pre-revolutionary Andover, Massachusetts, one deed of gift gave the family homestead to an unmarried son when he reached 30 but required him to "*take ye sole care of his father Henry Holt and of his natural mother Sarah Holt*" for the rest of their days and to provide for all their needs, which were carefully detailed. Failure to do so would result in forfeiture of the property. One Joseph Winslow left all his movable properties to his wife for her to distribute after death in accordance with their offspring's performance of filial duties. Such precautions would hardly have been necessary unless this mistreatment of elders was common. The threat of revocation was a widespread and doubtless prudent retention of economic power.

Economically-based parental control of adult children often resulted in conflict. One eighteenth century son, Robert Carter, remained under his father's authority for 20 years. Bitter conflicts bordering on violence ensued as a result of this continued economic dependence. One evening Robert invited friends over for an evening of gambling. The father, Landon Carter peremptorily ordered the cards and tables removed. He later recorded in his diary that: "*I was told by a forty-year old man he was not a child to be controulled.*" Fearing for his life, he armed himself with a pistol and inscribed in his diary:

> . . . *Surely it is happy our laws prevent parricide or the devil that moves to this treatment would move to put his father out of the way. Good God, that such a monster is descended from my loin* (cited by Fisher, 1977:75).

The 19th century family was also not immune from severed intergenerational ties. In 1889, a bill was presented to the board of commissioners of Brown County, Minnesota for the boarding of a poor, sickly old man who had been driven off by a son who no longer wanted to support him.

While the examples presented above do not represent a scientific survey of abuse and maltreatment of the aged, they do suggest that excellent care of the elderly by their children was not automatically provided and that the nostalgic picture of devoted, loving care provided by one's children may be little more than a desired but rarely attained myth.

Contemporary Family Relations

The relatively long period of (post maturity) child dependency and increasingly longer periods of elders' dependency on adult children are among the many circumstances of family life today that simply did not exist in the past.

Although the myth of the warm, close-knit, multigenerational families still exists, the evidence to support this myth is lacking. The reality is that, in spite of the media attention and increased resources given to protect and support children and elderly who are battered by family members, the quality and quantity of care provided today is far superior to that provided in earlier eras.

Abusive parents. Several national studies have documented the large amount of abuse perpetrated on children by their parents. Although children of all ages are victims, the very young and the young adolescent appear to be most at risk. A study based on interviews with 100 mothers of infants who were using outpatient clinics in Los Angeles found that almost half the mothers spanked infants under 12 months of age, and one-fourth had started spanking before the infant was six months old (Korsch, Christian, Gozzi, & Carlson, 1965). Wittenberg (1971), using parents representing a wide range of socioeconomic statuses, found no class differences in the methods used to discipline children. He did observe, however that 41% of the parents had used some form of physical punishment on their babies who were less than 6 months of age and 87% had used physical force before their children were two.

Children in the early adolescent years are also likely to experience high levels of physical and sexual abuse. The American Humane Association (1982) found that 36% of reported abuse cases involved children between 10 and 18; Gil (1968) found nearly 17% of abused youth were over 12. Knopoka (1975) conducted a nationwide study of adolescent girls and found that 12% had been beaten and 9% had been raped. Straus, Gelles and Steinmetz (1980) reported that 66% of youth 10-14 years old were struck and 34% of those 15-17 had been hit.

A paradox facing researchers, social service personnel and policy makers is that although the public is increasingly concerned over severe abuse of children, and these levels are beginning to decline, corporal punishment is still considered to be an *"acceptable"* form of family violence. American parents are expected to mold and control their children, and they are given considerable freedom in selecting the mechanism used to obtain this control. This is quite different than the perspective adopted in Sweden where a parent can be imprisoned for a month for striking a child (Ziegert, 1983).

Several researchers have asserted that we need to be concerned about the high levels of physical punishment of children because it represents one end of a continuum, the other end of which is child abuse (Gelles, 1973; 1974; Gil, 1970; Steinmetz, 1977a; Straus et al., 1980). With increasing stress, behaviors initiated as physical discipline can easily become child abuse as the boundaries of parental control are dismantled.

Probably the most accurate data for estimating child abuse is the recent second national survey (Straus & Gelles, 1986) of parents' reports of specific "disciplinary" techniques. As shown in Table 1, nearly 31% of the parents pushed, shoved or grabbed a child, 55% slapped and spanked, 10% hit or tried to hit with an object and just over one half percent beat-up a child.

Even the most severe forms of parent-child violence are not single events. They occur periodically and even regularly in the families where these types of violence are used. Straus et al., (1980) found that 8% of the children had been kicked, bitten, or punched, (an average of nine times a year), 4% were beaten, (an average of six times a year), and 3% were abused with guns or knives. If a beating is considered an indication of "child abuse," then child abuse appears to be a chronic condition for many children, not a rare experience for few.

Table 1.   A Comparison of the 1975 and 1985 National Survey of Family Violence[a]

Items	Parent-to-Child	
	1975 (N=1,146)	1985 (N=1,428)
Threw something	5.4	2.7
Pushed/grabbed/shoved	31.8	30.7
Slapped/spanked	52.2	54.9
Kicked/bit/hit with fist	3.2	1.3
Hit/tried to hit/object	13.4	9.7
Beat up	1.3	0.6
Threatened/ gun or knife	0.1	0.2
Used gun or knife	0.1	0.2

[a]The 1975 data from Straus et al., 1980; the 1985 data from Straus & Gelles, 1986.

Abusive adult children. A recurring question revolves around the relationship between a parent's abuse of a child and the child's later use of violence on the parent. While the probability of attacks on non-violent parents is about 1 in 400, the likelihood of attacks on parents who used violence on their adolescent children increases to about 200 out of 400. Unfortunately, these unresolved parent-child conflicts and abusive family interaction continue throughout the life cycle.

In-depth interviews with 104 family caregivers who were caring for 119 elders found that 41% were verbally abusive, 17% forced food or medicine, 13% were psychologically abusive, and 12% were physically abusive (Steinmetz, 1988). Overall, about 23% (nearly 1 in 4 caregivers) used some form of abuse which could result in physical harm (See Table 2).

Table 2.   Control Maintenance Techniques Used by
          Caregivers and Elders

CMT	Percent	
	Caregiver (N=119)	Elder (N=104)
Verbal Abuse	41	34
Pout	--	61
Manipulate	--	63
Cry	--	37
Psychological Abuse	13	--
Force Food/Medicine	17	--
Refuse Food/Medicine	--	24
Physical Abuse	12	18
Total Abuse	23	--

Source: *Duty Bound: Family Care and Elder Abuse.* Newbury Park, CA: Sage, 1988.

For the elder, violence becomes a method of last resort when there are no other means, such as prestige, authority, money, or independence to gain control. We should not be surprised that 34% of the elders were verbally abusive and 18% were physically abusive.

Although this sample of caregivers in many ways represented model caregivers, the abuse levels were high. Furthermore, they were intensified when the caregiver reported being stressed by having to perform caregiving tasks; general feelings of burden; or caring for an elder who was verbally and physically abusive. We can only imagine the levels of stress and burden and abuse that might occur in families that are forced into providing care because of a lack of alternatives.

RECIPROCAL OBLIGATIONS AND ABUSE

Intergenerational Comparisons

The findings discussed above might, on first glance, suggest that families have become more abusive and less concerned about children and elders in contemporary society.   However, several strands of evidence provide documentation that our treatment of and concern over elders and children has considerably improved in recent years.

First, a comparison of the data from the 1975 national survey of family violence with that collected in a second survey in 1985 substantiates a considerable decline for most child abuse behaviors. The severe violence index, composed of the scores for kicked/bit/hit with fist, hit/ threaten to hit with something, beat-up, threatened with gun or knife, and used a gun or knife was nearly 24% lower in 1985. For the most severe acts (all but the item "hit or tried to hit with something"), a 47% reduction was observed. Since this index is based on acts that are most likely to result in severe injury, a 47% decrease in child abuse confirms the considerable success of public awareness, education, service and intervention programs aimed at reducing child abuse.

Second, societal concern is no longer limited to the impact of sexual abuse or severe physical abuse on children, but also the impact of less severe physical and non-physical abusive acts.   The current child abuse campaign, for example, emphasizes the abusive nature of verbally berating a child.

Third, and most important, is that in a relatively short time behaviors such as strapping a child, that were once considered to be acceptable disciplinary measures would now be defined, by law, as abuse.   This suggests that in spite of the rhetoric of those spouting the Biblical injunction that we should not spare the rod, our society will no longer tolerate abusive treatment of children.

Similarly, the media has drawn our attention to the plight of the battered elderly.   However, self abuse, self neglect and emotional abuse constitutes the majority of cases handled by adult protective service workers.   Elders cast aside by their family and left to their own resources, such as described in the cases from earlier centuries, are relatively rare.

Our concern over the need to provide services to the large number of families who are caring for an elder can be measured by the rapid growth in the number of programs sponsored by state and federal governments, public and private and religious agencies, and corporations.

Intergenerational Transmission

Regardless of the support services provided by non-family entities, it is clear that families still provide most of the care for children and elderly parents. The parent-child bond endures in a variety of reciprocal and interacting ways over the life cycle. Therefore, the intergenerational transmission of values, attitudes and behaviors regarding the caregiving role throughout the life cycle is a critical link in explaining behavior. As a 61-year-old son who had been caring for his 92-year-old mother for over 15 years and his 89-year-old mother-in-law for 9 years noted:

> *My wife's mother took care of her own. They moved to their parent's home and stayed with them for a period of time. My mother's mother died in childbirth, her father died at sea, and an aunt brought her up. Later she took that aunt in. Following that she took in another lady that had taken care of my aunt. So she had two elderly people that she took care of until they died. I guess I hadn't thought of that before* (Steinmetz, 1988).

The impact of child abuse. Extensive studies document the impact of child abuse on the child's own life, the lives of other family members, and on the society at large. Numerous researchers have found a link between child abuse and parricide (Bender, 1959; Corder *et al.,* 1976; Sadoff, 1971; Sargent, 1962; Tanay, 1975). The link between an abusive childhood and spouse abuse has also been carefully documented (Gayford, 1975; Steinmetz, 1987; Straus *et al.,* 1980; Walker, 1984).

While an abused child is more likely to attack members of his or her own family, the violence extends beyond the family. Climent and Ervin (1972) found that assault by the mother or father was a common factor in the backgrounds of his clinical sample of violent individuals, while this factor was relatively rare in the control group.   In another study, murderers were more likely to have suffered severe physical beatings and traumatic incidents than their brothers who served as a control group (Palmer, 1962).

A brutalizing childhood was found to be characteristic of rapists (Brownmiller, 1975; Hartogs, 1951), individuals with split personalities (Schreiber, 1973), those who kill (Bender & Curran, 1940; MacDonald, 1967; Palmer, 1962), and those who attempt suicide (Connell, 1972; Jacobs, 1971). A study of the childhood environment of 14 political assassins revealed disorganized, broken families, parental abuse and rejection, and marginal integration into society (Steinmetz, 1973; 1977a).

The impact of elder abuse. Even in less violent forms, the use of physical force is passed from generation to generation

(Steinmetz, 1977a; 1977b; 1988). A daughter, 39, who has been caring for her parents for 5 years, clearly articulated the strength with which these patterns are transmitted.

My mother and I have a good fight now and then. . . Well, everybody does. You can't have a good relationship unless you have a good fight. . . If she said something to hurt me, I would come back with something just as bad. That's a normal reaction, If they hurt your feelings, you want them to hurt as bad . . . (Steinmetz, 1988).

As illustrated by the above quote, many of the abusive techniques used by elders parallelled those used by the caregivers. This is not surprising since it appears that these caregivers learned these ways of interacting during their own childhood. Now as middle-aged children and frail elderly parents, these patterns, established early in life, continue.

A comparison of the patterns of interaction between the elders and caregivers revealed that families in which the elder used verbal or physical abuse to gain control were the same families in which the caregiver used verbal or physical abuse. Likewise, caregivers who "ignored" the elder as their way of maintaining control, were caring for elders who "pouted," usually expressed by staying in their room or leaving the area where the rest of the family had gathered--a method quite similar to "ignoring" (Steinmetz, 1988).

Similarities and Differences in Responsibility and Obligation

There are several similarities between the battered child and the battered parent. Both the child and elderly parent are in a dependent position, relying on their caretaker for basic survival needs. Because the family's major role is fulfilling the expressive needs of its members, society expects caregivers to protect those who are dependent. Finally, both the dependent child and the dependent elder can be a source of emotional, physical and financial stress to the caregiver.

There are, however, major differences. While the cost of caring for one's children is a socially recognized responsibility and burden, the emotional and economic responsibility for the care of one's elderly parent over a prolonged period of time--a problem not likely to be faced by most families in the past--has not been acknowledged.

Although parents often provide care and support to their children while they are in college or during a crisis, their legal responsibility to provide this care ends when their children reach 18. There are no limits of liability for caring for one's parent, and proposed "relative responsibility" laws may force adult children to provide unlimited care after their parent's resources have been depleted (See

Steinmetz, 1988, for a fuller discussion of these laws and their
implications).

As Streib (1972) has suggested, studies on intergenerational
relationships among older families have tended to overlook the rights
and needs of adult children and to blame them if they have not
provided a comfortable, happy arrangement for their aged parents, a
situation that may last 10 to 15 years.

CONCLUSION

Parents have a clearly defined legal mandate to provide for the
care of their children. However our expectation to provide care for
our elderly kin is embedded in traditions and surrounded by myths.
While a societal mandate to provide care exists, a legal mandate
with resources and support systems that would enable the delivery
of this care without undue detriment to other family members and
themselves, is lacking. Adult children are the major support system
for their elderly kin and they are generous in their provision of this
care (Shanas, 1979; Sussman, 1959; 1965; Sussman & Burchinal,1962a).
Furthermore, this aid is part of a life-long pattern of mutually
beneficial, reciprocal intergenerational support (Hill, Foote, Aldous,
& Carlson, 1970; Sussman & Burchinal, 1962b; Sussman & Romeis,
1981). However, this intergenerational linkage is preserved not only
by a duty or obligation to provide support to the dependent at both
ends of the life span, but also by the emotional environment in
which this support is offered and received. A 66-year-old daughter
who has been providing care for her 86-year-old father for four
years summarized these feelings quite explicitly:

> I would say that if a parent has been a very loving
> parent and caring parent then a child could handle that,
> but when you know that they haven't been, it is very
> difficult to handle, very difficult. It's said that you
> have to forget what is in the past. Honey, you don't
> forget these things. Even an old dog, if you have
> beaten him, will cringe when he sees you. Even he
> doesn't forget. He either cringes or bites you.   .   .
> The thing that irks me so is that when I was little and
> needed he was never there. **He didn't take care of
> me!** . . . I've never heard him say thank-you (Steinmetz,
> 1988).

As Sussman (1959) first noted, there are intergenerational ties
that link parent to child and child to parent throughout the life
span. It is important, however, that we differentiate between the
provision of care and the quality of care that is provided. As our
society becomes more complex and the demands of multiple roles re-

sult in increasing levels of stress, we may find that the family is no longer able to adequately fulfill the expressive role without additional societal supports. ▓

Initial support for this research was provided by a grant from the State of Delaware, Division of Aging. Partial support for the preparation of this paper was provided by a grant from the Administration of Aging #90-AM-0204.

REFERENCES

American Humane Association. *National analysis of official child and neglect and abuse reporting.* Denver: CO, 1982.

Bates, R.P. *Child abuse-the problem.* Paper presented at the 2nd world conference of the International Society on Family Law. Montreal, June, 1977.

Bender, L. Children and adolescents who have killed. *American Journal of Psychiatry,* 1959, 116, 510-513.

Bender, L., & Curran, F.J. Children and adolescents who kill. *Criminal Psychology,* 1940, 3(4), 297-322.

Bremner, R.H. *Children and youth in America: A documentary history* (vol. 1). Boston: Harvard University Press, 1970.

Brody, S. J. The family caring unit: A major consideration in the long term support system. *The Gerontologist,* 1978, 18(Dec.), 556-561.

Brownmiller, S. *Against our will: Men, women and rape.* New York: Simon and Schuster. 1975.

Climent, C.F., & Ervin, F.R. Historical data in the evaluation of violent subjects. *Archives of General Psychiatry,* 1972, 27, 621-624.

Connell, P.H. Suicidal attempts in childhood and adolescence. In J.G. Howells (Ed.), *Modern perspectives in child psychiatry.* New York: Brunner/Mazel, 1972.

Corder, B.F., Ball, B.C., Haizlip, T.M., Rollins, R., & Beaumont, R. Adolescent parricide: A comparison with other adolescent murder. *American Journal of Psychiatry,* 1976, 133(Aug. 8), 957-961.

deMause, L. (Ed.). *The history of childhood.* New York: The Psychotherapy Press, 1974.

Fischer, D.H. *Growing old in America.* New York: Oxford University Press, 1977.

Gayford, J. Wifebattering: A preliminary survey of 100 cases. *British Medical Journal,* 1975, 1, 195-197.

Gelles, R.J. Child abuse as psychopathology: A sociological critique and reformulation. *American Journal of Orthopsychiatry,* 1973, 43, 611-621.

180   Parental and Filial Relationships

Gelles, R. J. *The violent home: A study of physical aggression between husbands and wives.* Beverly Hills, CA: Sage, 1974.

Gil, D. Violence against children. *Journal of Marriage and the Family*, 1968, 33(4), 637-648.

Gil, D. *Violence against children: Physical child abuse in the United States.* Cambridge: Harvard University Press, 1970.

Grevan, P. *Four generations: Population, land, and family in colonial Andover, Massachusettes.* Ithaca, NY: Cornell University Press, 1970.

Haber, C. *Beyond sixty-five.* New York: Cambridge University Press, 1983.

Hareven, T. K. Family time-historical time. *Daedaulus*, 1977, 106 (Summer), 57-71.

Hartogs, R. Discipline in the early life of six delinquent and sex criminals. *Nervous Child*, 1951, 9(March), 167-173.

Hill, R., Foote, N., Aldous, J., Carlson, R., & MacDonald R. *Family development in three generations.* Cambridge, MA: Schenkman Publishing Co., 1970.

Jacobs, J. *Adolescent suicide.* New York: Wiley Interscience, 1971.

Kent D. P. Aging: Fact and fancy. *The Gerontologist*, 1965, 5 (June), 51-56.

Knopoka, G. *Young girls: A portrait of adolescence.* Englewood Cliffs: Prentice-Hall, 1975.

Korsch, B.M., Christian, J.B., Gozzi, E.K., & Carlson, P.V. Infant care and punishment: A pilot study. *American Journal of Public Health*, 1965, 55, 1880-1888.

MacDonald, J.M. Homicidal threats. *American Journal of Psychiatry*, 1967, 124, 475-482.

Mather, Cotton. *A brief essay on the glory of aged piety.* Boston: Kneeland & T. Green, 1726.

Mather, Cotton. *Dignity and Duty.*

Mather, Increase. *Two discourses.* Boston: B. Green, 1716.

Nursing Home Care in the United States: Failure in Public Policy, Introductory Report. Senate Report, 93-1420, 93rd. Cong. 2d Session. 16 (1974). [(Cited in The *Nursing Home Law Letter*, 56 (Feb.) 1, 1982.]

Palmer, S. *The psychology of murder.* New York: Thomas Y. Crowell, 1962.

Radbill, S. X. A history of child abuse and infanticide. In R.E. Helfer & C. H. Kempe (Eds.), *The battered child* (1st Ed.) Chicago: University of Chicago Press, 1968.

Records of the Boston Selectmen. A Report of the Record. Commissioner of the City of Boston: Vol 15, 1736-1742. Boston: Rockwell & Churchill City Printers, 1936.

Sadoff, R.L. Clinical observations on parricide. *Psychiatric Quarterly*, 1971, 45, 65-69.

Sargent, D.A. Children who kill: A family conspiracy. *Social Work*, 1962, 7, 35-42.

Shanas, E.   The family as a social support system in old age.   *The Gerontologist*, 1979, 19(April), 169-174.

Schreiber, F.R. *Sybil.* New York: Warner Books, 1973.

Smith, D.B.   *Inside the great house: Planter family life in the 18th century Chesapeake society.*   Ithaca, NY: Cornell University Press, 1980.

Steinmetz, S.K.   Family backgrounds of political assassins.   Paper presented to the American Orthopsychiatric Association, 1973. (Reviewed in *Human Behavior*, January, 1974)

Steinmetz, S.K.   *The cycle of violence: Assertive, aggressive and abusive family interaction.*   Praeger, New York, 1977. (a)

Steinmetz, S.K.   The use of force for resolving family conflicts: The training ground for abuse.   *Family Coordinator*, 1977, 33(4), 19-26. (b)

Steinmetz, S.K.   Family violence: Past, present, and future.   In M.B. Sussman & S.K. Steinmetz (Eds.), *Handbook of marriage and the family.*   New York: Plenum Publishers, 1987.

Steinmetz, S.K.   *Duty bound: Family care and elder abuse.*   Beverly Hills, CA: Sage Publication, 1988.

Straus, M.A.   Family patterns and child abuse in a nationally representative American sample.   *Child Abuse and Neglect*, 1979, 3, 213-215.

Straus, M.A., Gelles, R.J., & Steinmetz, S.K.   *Behind closed doors: Violence in the American family.*   New York: Doubleday, 1980.

Straus, M.A., & Gelles, R.J.   Societal change and change in family violence from 1975 to 1985 as revealed by two national surveys.   *Journal of Marriage and the Family*, 1986, 48(August), 465-479.

Streib, G.F.   Older families and their troubles: Familial and societal responses.   *Family Coordinator*, 1972, 21(Feb.), 5-19.

Sussman, M.B.   The isolated nuclear family: Fact or fiction.   *Social Problems*, 1959, 6(Spring), 33-340.

Sussman, M.B.   Relationships of adult children with their parents in the United States.   In E. Shanas & G. Streib (Eds.), *The older person in the family: Challenges and conflicts.*   Iowa City: University of Iowa Institute of Gerontology, 1965.

Sussman, M.B.   (Burgess address) Law and legal systems: The family connection.   *Journal of Marriage and the Family*, 1983, 45(Feb.), 9-21.

Sussman, M.B., & Burchinal, L.   Kin family networks: Unheralded structure in current conceptualizations of family functioning.   *Marriage and Family Living*, 1962, 24(August), 231-240. (a)

Sussman, M.B., & Burchinal, L.   Parental aid to married children: Implications for family functioning. *Marriage and Family Living*, 1962, 24(Nov.), 320-332. (b)

Sussman, M.B., & Romeis, J.C.   Family supports for the aged: A comparison of U.S. and Japan responses.   *Journal of Comparative Family Studies*, 12, 4(autumn), 475-492.

Tanay, E. Reactive parricide. *Journal of Forensic Sciences*, 1975, 2(1), 76-82.

Walker, L.E. *The battered woman syndrome.*   New York: Springer, 1984.

Wittenburg, C. Studies of child abuse and infant accidents. In J. Segal (Ed.), *The Mental health of the child: Program of the National Institute of Mental Health.*   Washington, D.C.: U.S. Government Printing Office, 1971.

Ziegert, K.A. The Swedish prohibition of corporal punishment: A preliminary report. *Journal of Marriage and the Family*, 1983, 45(Nov.), 917-926.

☒ 14 ☒

MOTHER-DAUGHTER BONDS IN THE LATER YEARS: TRANSFORMATION OF THE 'HELP PATTERN'

Doris Y. Wilkinson

INTRODUCTION

Despite the reduced familism of the middle-class family, parents and their married children . . . have a pattern of giving and receiving between them. Moreover, this pattern is related to the continuity of intergenerational family relationships (Sussman, 1953b:27).

While observing an intensive care unit of a local hospital with my sister recently, I wondered: "Why are most of the patients female and most of those waiting adult women?" I later learned that those in intensive care were stroke or seizure victims - women in their late 60s to early 70s. Those waiting in the corridors were older daughters - single, married, or divorced. Although the backgrounds of these women were dissimilar with respect to class, race, rural versus urban residence, and even cultural exposures, for a brief moment in their lives, a common experience was shared. Each had a seriously ill mother, and the daughters, functioning as the primary overseers, were solely responsible for communicating with hospital personnel and for providing emotional support. This situational similarity in role performance occurred in spite of the unique intricacies of their past family environments. What is the nature of this enduring support bond between mothers and daughters which crystallizes sex role distinctions and discloses the gender factor in society?

For many years, the study of mother-daughter interaction, with a special emphasis on personality and relational conflicts during adolescence, has preoccupied researchers (See Wilkinson, 1988b for an extensive bibliography on this material).   However, Neisser, (1967:90) has noted that:

> *When a woman finds herself continually at odds with her daughters, the fault may lie not so much in their behavior as in the lack of satisfaction she derives from her own life.*

This notion is pertinent to understanding aspects of the quality and form of the mother-daughter bond during the family's later years.   A number of parent-offspring relational and adjustment problems which were of interest thirty to forty years ago persist in family research.   Representative of the prevailing topics are those which center on children of divorce, maternal overprotection, maternal employment, adopted children, and the self concepts and behaviors of those from interracial and inter-ethnic marriages.   Of particular interest in child development theory has been how various aspects of the mother's role performance contribute to the *"life adjustment of the child"* (Thomas, 1955:1).   With respect to this, it has been aptly observed that:

> *A girl's attitude toward being a woman, her aspirations, tastes and feelings are similarly influenced by the pursuits her mother follows and how her mother feels about that life style* (Neisser, 1967:88).

Further, reactions to teachers and even learning ability mirror early socialization and mother-daughter interaction.   As Neisser suggests:

> *If her experience with authority, in the person of her mother, has not been alarming, she will be predisposed to accept teachers as potentially friendly.   She will tend to regard mastering subject matter as within the range of her powers if her intellectual curiosity has not been stifled and if winning approval has not been difficult.   If making a mistake has not been overwhelmingly humiliating and inquiries and experiments have been encouraged, she will tend to be a good learner* (1967:65).

The attitudinal and behavioral consequences which emanate from this formative conditioning permeate the quality of future mother-daughter relations (Booth, 1984; Glenn & McLanahan, 1981; Rempel, 1985).

In the 1980s, the family literature has centered on a plethora of social, legal, and ethical concerns such as surrogate parenting (Harrison, 1987), mothers of abused offspring, homeless women, homeless children (Bassuk & Rubin, 1987), and individuals without kinship ties.   While these are critically important areas for further

study, there is very little policy oriented research which con
trates on adult daughters and the multiple role expectations held
them by family members, their communities, and the larger society.
Data are limited on the diverse and long-term obligations they fulfill
regardless of socioeconomic position, marital and employment status,
or ethnicity (Brody, Johnsen, & Fulcomer, 1984; Brody & Schoonover,
1986).

Further, there is little consistent information regarding the
impact of filial obligations on personal ambitions, career goals,
performance in the workplace, life styles, and even life chances
(Archbold, 1980; 1983; Brody & Lang, 1982; Seelbach, 1978; Seelbach
& Sauer, 1977; Soldo & Mylloyma, 1983). What we do know is that
assistance to aging and/or ill parents and intergenerational bonding
have been perpetuated in spite of the above factors. It is also
known that gender distinctions are constant in the enactment of
filial expectations (Horowitz, 1985).

Sociological and behavioral science studies centering on role
identification and the emotional ties between mothers and daughters
have underscored the traditional function of grandmothers in their
daughter's "mothering" behavior, expectations for older family members in rural settings, and the structure of family interaction in
later years. The sustained link between mothers and daughters has
been a primary focus (Wilkinson, 1988b). Recent popular writings,
however, have appeared to depict the relationship as neither mutually rewarding nor emotionally healthy (Crane, 1988; Crawford,
1984; Davis, 1986; Hyman; 1985). It is instructive to keep such
accounts in mind as the transformation of the mother-daughter bond
is examined within the context of the dynamics of sociocultural
changes.

FILIAL RESPONSIBILITY IN THE MODERN AGE

There has been a relative neglect of the multiple and time-
based dimensions of filial role responsibilities in a socio-political
milieu with an emergent ethos which appears to contradict generational bonding (Suitor, 1987; Troll, Miller, & Atchley, 1979). Recently, however, considerable attention has been devoted to role
reversal and patterns of interaction between elderly parents and
female adult offspring (See Wilkinson, 1988b for an extensive bibliography). What the data have shown is that there may be no
other status-role set within the family, regardless of its economic
position, that is characterized by the array of expectations and
obligations as that of the daughter, an institutionalized occurrence
which is apparent across ethnic boundaries. Cross-culturally, from
adolescence to adulthood, the configuration of expressive and instrumental roles which daughters enact covers a vast range, including,
in the later years, caregiver, in some form, for offspring, spouses,
as well as both aging parents. With respect to this, Sussman aptly

noted in the late 1970s that "*this generation, in addition to caring for the young, now has to care for old parents*" (1976:233).

This descriptive synthesis identifies both the perpetuation of and modifications in the 'help pattern' in the later stages of the family life cycle, especially as parents become functionally dependent. The focus centers on the mother-daughter dyad. Notions of linkages across the generations, as manifested in reciprocal aid and hence in the fundamental components of the 'help pattern,' are reappraised. The discussion emphasizes a subject which has not gained the attention it deserves: the transformation of the link between mothers and daughters as the family experiences its "passages." The concepts and principles advanced are applicable to mothers and sons, fathers and daughters, and especially to mothers given the increased life expectancy of older women and the consequent likelihood of sickness and functional dependency. The aim is to extend interpretations of intergenerational connectedness by concentrating on daughters as the primary caregivers for elderly and functionally dependent mothers, to highlight contemporary social realities and policy concerns emanating from the reconstruction of the 'help pattern' as parents and offspring age, and to suggest new research avenues (Fischer, 1981; Gibson, 1972; Kessin, 1971; Kulis, 1987; Popenoe, 1987; Stevens & Boyd, 1980; Wood & Robertson, 1978).

Filial duties necessitated by early retirement, the growth in the elderly population, widowhood, parental disability, and ill health are emergent social issues with significant family and policy ramifications (Rix, 1984; Wilkinson, 1987b). Among others, these include compensation for unpaid labor and the reformulation of private and public agency support services. Thus, on the threshold of the 21st century, with a growing elderly population and escalating medical care costs, researchers and decision makers have not delineated which institutions might assist young adult and middle-aged women in the support of elderly or functionally dependent parents.

While interest in the mother-daughter union is not recent, there is an absence of analytic frames of reference and practical evaluations of the fundamental changes in the character of intergenerational succession as the family traverses the life cycle. There have been few reality based scientific analyses of the pervasive impact of social and cultural transitions on filial responsibilities and familistic sentiments. In addition, there has been no probing examination of the stresses associated with filial obligations in a context which has generated for women new value and role definitions as well as emergent aspirations and needs. In earlier research, the inculcation of values and conditioning were translated with respect to internalized role orientations, the sense of security, and subsequent parental identification which the daughter acquired from the mother (Baruch, 1972). Yet, in the contemporary political economy, modified definitions of role expectations and enactments have evolved.

THE SUSSMAN 'HELP PATTERN' REVISITED

Thirty-five years ago Marvin Sussman offered an insightful and informative analysis of American middle class behavior. His research on urban families concentrated on two interrelated topics which were not in vogue at the time: the myth of the isolated nuclear unit   and the prevalence of a 'help pattern' (Sussman, 1953b). He called attention to whether prevailing notions were supported by empirical research. His study of family and kinship relationships led to the conclusion that intergenerational ties among kin have "*far more significance than we have been led to believe in the life processes of the urban family*" (Sussman, 1959:339). Differences in the help offered, particularly financial aid and the types of services exchanged, were found to vary by socioeconomic status. Sussman observed that the class distinctions were more indicative of life style variations than an unwillingness to "participate in the mutual aid network." Typically, working class family members lacked the economic resources to offer assistance.

In Sussman's original appraisal of the 'help pattern' as an indicator of generational linkages, the concept of mutual aid was crystallized. While support during illness remained the primary form, distance influenced the frequency of contact and the nature of aid offered (Litwak, 1960; Sussman, 1976). Yet, the study of middle-class families accented the prominence of familial reciprocity in times of need. The data demonstrated that urban families tended to live in close physical proximity to one another and were involved in a successive pattern of mutual assistance. A serendipitous consequence of the 'help pattern' was the maintenance of social exchange and friendly associations among parents, offspring, and extended kin (Sussman, 1953b). Of particular interest at the time was the support provided by parents to their daughters.

> *Parents gave the greatest amount of nursing care during the confinements of their daughters, either by taking the older grandchildren to live with them or by performing nursing services. As might be anticipated, they were asked to help by their daughters more frequently than by daughters-in-law* (Sussman, 1953b:25).

In summary, several important findings emerged from Sussman's original perceptive analysis of kinship networks and intergenerational ties. These included:

▓ A willingness of relatives to assist each other,

▓ An expectation on the part of children that parents would provide encouragement and aid at the onset of marriage,

 ▓ The substitution of service for money,

 ▓ A norm of reciprocity between parents and their married
   children, and

 ▓ A shared sentiment of feeling free to seek help from each
   other.

These results established a foundation for present assessments of the
mother-daughter bond in a sociocultural milieu which is transforming
its character.

The Impact of Social Change

For all ethnic groups in the United States, social, cultural, and
technological changes have permeated and modified the structure and
functioning of the family (Wilkinson, 1984; 1987a).   Contemporary
American familial norms and relationships thus reflect the dramatic
shifts in societal beliefs, values, and behaviors.   Among the environmental events which have affected the nature of family interaction
and intergenerational continuity are: a fluctuating economy, rapid
cultural transitions, biomedical innovations, and adjustments in
Medicare and in other health insurance programs (Wilkinson, 1987b).
Specific societal conditions bearing directly on the family constellation and its internal dynamics are innumerable: the opening of the
opportunity structure, larger numbers of employed women, an increase in the numbers of professional women, growth in dual-career
families (Rapoport and Rapoport, 1976), an increase in the frequency
of long distance marital arrangements, high divorce rate, changing
sexual mores, and a growing population of older persons, especially
women.   Juxtaposed with socioeconomic differentials, the Women's
Liberation Movement has had a profound impact in restructuring
women's personal needs, perceptions, and definitions of conventional
role expectations.   All of these have contributed to readjustments in
family organization, parent-offspring relationships, spousal attitudes
and expectations, and role enactments.
Interestingly, with the changes in family norms and relationships, there has been some assumption of traditional female responsibilities by fathers such as housework and caring for children
(Coverman & Sheley, 1986; Hartmann, 1981; Hellmich, 1988).   However, this modern "role playing" has been virtually inconsequential
and mothers and grandmothers remain the primary care agents and
significant molders of the attitudes and behaviors of their sons and
daughters (Cohler, 1981; Sommerville, 1972; Thomas, 1955; von
Hentig, 1946; Wilkinson, 1988a).   In addition to the personality and
values of the mother affecting the character and direction of daughters' and sons' behaviors, her marital and employment status have
also been associated with offsprings' future role perceptions and

enactments (Stevens & Boyd, 1980).   Each of the factors noted above has consequences for the content and form of intergenerational relationships.

Caregiving: Its Social and Psychological Costs

In view of the complexity of social environmental effects on the fabric of family life and the changing roles and needs of women, a more thorough scrutiny of generational ties encompassed by the 'help pattern' is warranted. Although there is limited literature on the reconstruction of its multiple dimensions including the social, economic, and psychological costs of the caregiving role (Cantor, 1981; Kessin, 1971; Poulshock & Deimling, 1984; Robinson, 1983; Seelbach & Sauer, 1977; Simon, 1986; Zarit, 1980), several important views have been advanced for future study and for consideration in policy formulation.

Although filial expectations are undergoing continual conversion and redefinition, their female-based form illuminates the entrenchment of traditional sex-role norms and hence the gender factor in society.   Current data indicate that adult daughters are primarily responsible for providing emotional, financial, and supplemental health care to elderly parents, especially their mothers. With respect to this, Sussman (1976:227) noted in his article, *"The Family Life of Old People,"* that regardless of geographic distance, the care of kin "is largely the work of female members." For both young adult and middle aged daughters, regardless of marital status, there are varied consequences which result from the enactment of multiple roles.   Understandably, role reversal *"requires new learning and places upon the middle generation potentially anxiety-provoking responsibilities"* (Sussman, 1976:233).   When the severity, extent, and impact of parental aging, illness, or other forms of functional dependency are taken into account, measurable strains are associated with caregiving tasks. Heightened expectations and levels of aspiration contribute to these strains.   In one study of fifty elderly never-married women from metropolitan areas in the northeastern part of the country, 84% were found to have served as the primary caregivers for their parents.   While satisfaction was derived from the helping role, a prominent factor in intergenerational contact, personal and financial pressures was encountered (Simon, 1986).

Moreover, if a husband or father is elderly or deceased or when other siblings do not share equally in providing assistance to parents in need, stresses multiply for the family member who becomes the primary caregiver.   When adult daughters serve in this capacity for extended periods of time, they may incur physical, psychological, and social costs.   Simon found that physical costs can range from fatigue to serious illness.   In her study, the latter were manifested in a variety of symptoms such as back problems, resulting from lifting parents needing such assistance, and high blood pressure. Over extended periods of time, feelings of bitterness, anger, and resentment emerged. Further, some of the adult daughters experienced depression and felt a sense of hopelessness. Such emotional responses were not merely the outcome of the enactment of role expectations but emanated from the suppression of feelings (Simon, 1986). Simon also discovered that the women in her study did not perceive their illnesses as being directly associated with caregiving tasks. Rather, *"they understood their physical difficulties to be exacerbated by the volume of work and worry that caring for an aging parent entails"* (Simon, 1986:35).

Stresses encountered by daughters as primary caregivers are associated not only with the type, severity, and duration of parental disability or illness but with the parent's emotional state and coping capabilities. These stresses include work related difficulties, social isolation, as well as financial pressures. Obvious financial costs are incurred for medicines, hospital services, health care equipment, and doctor's fees not covered by Medicare or supplemental insurance plans. Social costs may cover a wide range from loss of employment opportunities to estrangement from friends. Simon discovered that personal ambitions and goals can be thwarted by the career restricting impact of the caregiving role. For some of the women in her sample, *"jobs and careers were changed, redirected, geographically relocated, made part time. . ."* (1986: 37).

CONCLUSIONS

Until the probing analyses of Sussman questioned the validity of the 'atomistic family' construct (Sussman, 1953b; 1954; 1959; 1960; 1965b), conceptions of a detached or isolated nuclear unit predominated in the sociological literature. Although his informative work appeared more than three decades ago, presently in the behavioral sciences and in the policy sphere, there appears to be no application of the cumulative data on the multidimensional nature of the caregiving role of offspring, especially that performed by adult daughters throughout the family life cycle. There is a need for systematic measures of the extent, type, and impact of their filial responsibilities. It is important to note that:

> Today's elderly population was socialized at a time when family relationships outweighed individual goals and feelings; this may have resulted in a strong sense that children should be obligated to care for aging parents (Johnson, 1981-82: 272).

The continuity of relations with elderly parents, in spite of occupational or geographic mobility (Litwak, 1960; Sussman, 1976: 229), can no longer be limited to an examination from the perspective of learned behavior or traditional values and expectations. Research must take into account the diverse personal outcomes and social "costs" for adult daughters in a changing cultural milieu wherein women continue to be expected to enact diverse sets of roles (e.g., wives, mothers, householders, workers, kinship liaisons, and/or parent substitutes). The measurable effects of long-term filial obligations for aged and/or functionally dependent parents are not only manifested in life style alterations but in the modification of personal needs and goals. Concomitantly, as some of the literature has indicated, when women serve as the primary caregivers and assume multiple roles, potential health effects may result from the associated tasks as well as from the suppression of needs, emotions, and aspirations. If health status is affected, predictable outcomes exist for their productivity as workers and hence for their life chances.

The changes in traditional filial expectations warrant intensive scrutiny as the female population traverses the life cycle and as single, divorced, or widowed daughters are expected to take on functions which strain their capacities and the quality of the bonding between them and their mothers.

There needs to be a realistic appraisal of alternative forms of support for aging, disabled, and/or functionally dependent family members. The possibilities for such support encompass an array of voluntary social groups and networks. These include:

* Friends and neighbors who contribute time and energy to caregiving tasks;

* Social agencies, clubs, and even sororities and fraternities that design constructive projects such as delivering needed services or arranging transportation to hospitals, doctors' offices, and clinics;

* Community kitchens which provide meals and even sitter arrangements;

* Senior citizens' organizations which serve as community centers and which are open at all times; and

* Churches that are socially conscious and service-oriented for their members and other community residents (See Arling, 1976; Breslau & Haug, 1972; Cantor, 1979; Taylor & Chatters, 1986).

The pooling of community resources and the establishment of tax supported care centers for aging parents would supplement the multidimensional care provided by female offspring. It is imperative that various structural alternatives be explored and delineated as the elderly population increases, medical care costs escalate, social security and Medicare benefits diminish, limited family resources are exhausted, and young adult and middle-aged women attempt to maintain traditional roles.

In spite of the gradual evolution of divergent role contingencies and expectations as offspring and parents age, the enduring nature of the intergenerational link between daughters and mothers has been affirmed:

> . . .*almost nothing appears to diminish feelings of filial responsibility.   The pressures of life notwithstanding, the mother-daughter relationship provides both with that sense of historical continuity* . . . (Bromberg, 1983:90).

Sussman's earlier and recent works have enriched our understanding of the dynamics of this bond as it confronts inevitable transformation in a rapidly changing social order. ▓

REFERENCES

Arcbold, P.G.   The impact of caring for an ill elderly parent on the middle-aged offspring.   *Journal of Gerontological Nursing*, 1980, 6, 78-85.

Arling, G.   The elderly widow and her family, neighbors and friends. *Journal of Marriage and the Family*, 1976, 38,(November), 757-768.

Baruch, G.K.   Maternal influences upon college women's attitudes toward women and work. *Developmental Psychology*, 1972, 6, 32-37.

Bassuk, E., & Rubin, L.   Homeless children: A neglected population. *American Journal of Orthopsychiatry*, 1987, 57 (April), 279-286.

Booth, R.   Toward an understanding of loneliness.   *Social Work*, 1984, 28, 116-119.

Breslau, N., & Haug, M.R.   The elderly aid the elderly: The senior friends project. *Social Security Bulletin*, 1972, 35, 9-15.

Brody, E.M., & Lang, A.   They can't do it all: Aging daughters with aged mothers. *Generations*, 1982, 7, 18-20.

Brody, E.M., & Schoonover, C.B.   Patterns of parent care when adult daughters work and when they do not. *The Gerontologist*, 1986, 26, 372-381.

Brody, E.M., Johnsen, P.T., & Fulcomer, M.C..   What should adult children do for elderly parents?: Opinions and preferences of three generations of women. *Journal of Gerontology*, 1984, 39, 736-746.

Bromberg, E.M.   Mother-daughter relationships in later life: The effect of quality of relationship upon mutual aid.   *Journal of Gerontological Social Work*, 1983, 6, 75-92.

Cantor, M. Factors associated with strain among family, friends and neighbors caring for the elderly. Paper presented at the 34th annual meeting of the Gerontological Society of America, Toronto, 1981.

Cantor, M. Neighbors and friends: An overlooked resource in the informal support system. *Research on Aging*, 1979, 1, 434-463.

Cohler, B.J., & Grunebaum, H. *Mothers, grandmothers, and daughters*. New York: John Wiley & Sons, 1981.

Coverman, S., & Sheley, J. Change in men's housework and child-care time, 1965-1975. *Journal of Marriage and the Family*, 1986, 48(May), 413-422.

Crane, C. *Detour: A Hollywood story*. New York: Arbor House, 1988.

Crawford, C. *Mommie dearest*. N.Y.: G.P. Putnam's & Sons, 1984.

Davis, P., & Foster, M.S. *Home front*. New York: Crown, 1986.

Fischer, L.R. Transitions in the mother-daughter relationship. *Journal of Marriage and the Family*, 1981, 43, 613-622.

Gibbs, J.T. Identity and marginality: Issues in the treatment of biracial adolescents. *American Journal of Orthopsychiatry*, 1987, 57(April), 265-278.

Gibson, G. Kin family network: Overheralded structure in past conceptualizations of family functioning. *Journal of Marriage and the Family*, 1972, 34 (Feb.), 13-23.

Glenn, N.D., & S. McLanahan. The effects of offspring on the psychological well-being of older adults. *Journal of Marriage and the Family*, 1981, 43(May), 409-421.

Harrison, M., Social construction of Mary Beth Whitehead. *Gender & Society*, 1987, 1(Sept.), 300-311.

Hartmann, H.I. The family as the locus of gender, class and political struggle: The example of housework. *Signs*, 1981, 6, 366-394.

Hellmich, N. Men still don't do their part around the house. *USA Today*, 1988, (January 20), 1D.

Horowitz, A. Sons and daughters as caregivers to older parents: Differences in role performance and consequences. *The Gerontologist*, 1985, 25, 612-617.

Hyman, B.D. *My mother's keeper*. New York: Morrow, 1985.

Johnson, E.S. Role expectations and role realities of older Italian mothers and their daughters. *International Journal of Aging and Human Development*, 1981-82, 14, 271-276.

Kessin, K. Social and psychological consequences of intergenerational occupational mobility. *American Journal of Sociology*, 1971, 77, 1-18.

Kulis, S. Socially mobile daughters and sons of the elderly: Mobility effects within the family revisted. *Journal of Marriage and the Family*, 1987, 49 (May), 421-433.

Litwak, E., Geographic mobility and extended family cohesion. *American Sociological Review*, 1969, 25, 385-394.

Neisser, E.G. *Mothers and daughters: A lifelong relationship.* New York: Harper & Row Publishers, 1967, (revised edition).

Popenoe, D. Beyond the nuclear family: A statistical portrait of the changing family in Sweden. *Journal of Marriage and the Family*, 1987, 49(Feb.), 173-183.

Poulshock, S. W., & Deimling, G.T. Families caring for elders in residence: Issues in the measurement of burden. *Journal of Gerontology*, 1984, 39, 230-239.

Rapoport, R., & Rapoport, R.. *Dual-career families revisted: New integrations of work and family.* New York: Harper & Row, 1976.

Rempel, J. Childless elderly: What are they missing? *Journal of Marriage and the Family*, 1985, 47(May), 343-348.

Rix, S.E. Older women: The politics of aging. Washington, D.C.: Women's Research and Education Institute, Congressional Caucus on Women's Issues, 1984.

Robinson, B. Validation of a caregiver strain index. *Journal of Gerontology*, 1983, 39, 344-348.

Seelbach, W.C. Gender differences in expectations for filial responsibility. *The Gerontologist*, 1977, 17, 421-425.

Seelbach, W.C., & Sauer, W. Filial responsibility, expectations and morale among aged parents. *The Gerontologist*, 1977, 17, 492-499.

Simon, B.L. Never-married women as caregivers to elderly parents: Some costs and benefits. *AFFILIA Journal of Women and Social Work*, 1986, 1, 29-42.

Soldo, B.J., & Myllyuoma, J. Caregivers who live with dependent elderly. *The Gerontologist*, 1983, 23, 605-611.

Somerville, R. Future of family relationships in middle and older years: Clues in fiction. *Family Coordinator*, 1972, 21, 487-498.

Stevens, G., & Boyd, M. The importance of mother: Labor force participation and intergenerational mobility of women. *Social Forces*, 1980, 59, 186-189.

Stolz, L.M. Effects of maternal employment on children: Evidence from research. *Child Development*, 1960, 31, 749-782.

Suitor, J.J. Mother-daughter relations when married daughters return to school: Effects of status similarity. *Journal of Marriage and the Family*, 49, (May), 435-444.

Sussman, M.B. Parental participation in mate selection and its effect upon family continuity. *Social Forces*, 1953, 32, 76-81. (a)

Sussman, M.B. The help pattern in the middle class family. *American Sociological Review*, 1953, 18, 22-28. (b)

Sussman, M.B. Family continuity: Selective factors which affect relationships between families at generational levels. *Marriage and Family Living*, 1954, 16, 112-120.

Sussman, M.B. The isolated nuclear family: Fact or fiction. *Social Problems*, 1959, 6, 333-340.

Sussman, M.B. Intergenerational family relationships and social role changes in middle age. *Journal of Gerontology*, 1960, 15, 71-75.

Sussman, M.B. (Ed.).   American adolescents in the mid-sixties. *Journal of Marriage and the Family*, 1965, 27, 134-303. (a)

Sussman, M.B.   Relationships of adult children with their parents in the United States.   In E. Shanas &   G. Streib (Eds.), *Social structure and family generational relations.* Englewood Cliffs, New Jersey: Prentice-Hall, 1965. (b)

Sussman, M.B.   The family life of old people.   In R. Binstock & E. Shanas (Eds.), *Handbook of aging and the social sciences.*   New York: Van Nostrand Reinhold, 1976.

Taylor, R.J., & Chatters, L. M.   Church-based informal support among elderly blacks. *The Gerontologist*, 1986, 26, 637-642.

Thomas, R.C.   *Mother-daughter relationships and social behavior.* Washington, D.C.: The Catholic University of America Press, 1955.

Thompson, L., & Walker, A.J.   Mothers and daughters: Aid patterns and attachment. *Journal of Marriage and the Family*, 1984, 46, 313-222.

Troll, L.E., Miller, S.J., & Atchley, R.   *Families in later life.* Belmont, California: Wadsworth, 1979.

Von Hentig, H.   The social function of the grandmother. *Social Forces*, 1946, 24, 389-392.

Wilkinson, D.Y.   Afro-American women and their families. *Marriage and Family Review*, 1984, 7(Fall/Winter), 125-142.

Wilkinson, D.Y.   Ethnicity.   In M.B. Sussman & S.K. Steinmetz (Eds.), *Handbook of marriage and the family.* New York: Plenum, 1987. (a)

Wilkinson, D.Y.   Transforming national health policy: The significance of the stratification system. *The American Sociologist*, 1987, 18, (Summer), 140-145. (b)

Wilkinson, D.Y. Traditional medicine in American families: Reliance on the wisdom of elders. *Marriage and Family Review*, 1988, 11, 69-80. (a)

Wilkinson, D.Y.   Bibliography on mother-daughter relationships. Unpublished, 1988. (b)

Wood, V., & Robertson, J.F.   Friendship and kinship interaction: Differential effect on the morale of the elderly. *Journal of Marriage and the Family*, 1978, 40 (May), 367-375.

Zarit, S., Reever, K., & Bach-Peterson, J.   Relatives of the impaired elderly: Correlates of feelings of burden. *The Gerontologist*, 1980, 10, 649-655.

※ 15 ※

CROSS-GENDER RELATIONSHIPS AT WORK AND HOME OVER THE FAMILY LIFE CYCLE

Elina Haavio-Mannila

INTRODUCTION

The reciprocal linkages between the family and bureacratic institutions, especially the occupational setting, is a recurring theme (Hess & Sussman, 1984; Sussman, 1983; Sussman & Shanas, 1976). In this chapter cross-gender relationships at work and at home will be examined. These relationships are related to both structural factors, such as the extent of gender segregation of work and social status, and to normative factors, such as gender, marital, parental and age roles. Both factors regulate the behavior of husbands and wives, fathers and mothers across the life span.

Social relationships between men and women outside the marriage are regulated and restricted by social norms and taboos. These cultural restrictions contribute to a continuation of gender segregation in the labor market, gender inequality in the family, and informal cross-gender relationships at work.

The family life-cycle, a concept combining marital, parental and age status--a multifaceted indicator--and gender, are the major explanatory variables in this study. In order to assess the effect of gender roles of social interaction between men and women within the family setting as well as outside the family, the data have been analyzed separately for men and women throughout the study.

GENDER SEGREGATION OF WORK AND SOCIAL STATUS

Gender segregation of both paid and unpaid work is a universal phenomena, although the form and degree vary across cultures and time.   Functional gender segregation of work appears to be more common than physical separation of men and women in the workplace or at home.   Thus men and women conducting different sorts of work often have social contacts with each other (Kauppinen-Toropainen, Kandolin, & Haavio-Mannila, 1988; Reskin & Hartman, 1986).

In the industrialized countries the sexual division of domestic tasks at home has somewhat weakened as the employment of wives and mothers has increased.   Nevertheless, in most families the wife still carries the major responsibility for housework (Niemi, Kiiski, & Liikkanen, 1981; Pleck, 1984; Lewis & Sussman, 1986; Losh-Hesselbart, 1987).   At home the spouses have contact with each other even when they perform different household tasks, or share them unequally.   However, the home and the workplace are different social-sexual environments.   Thus, men and women are probably less sex-segregated at home than in the workplace.

In gender-integrated jobs informal cross-gender relations are more numerous and close than in gender-segregated jobs.   For example, cross-gender friendship and work romances are much more common in mixed work places, where men and women share work tasks, than in same sex workplaces (Haavio-Mannila, 1987a).

It also has been found that gender integration of paid work increases husband's participation in domestic work, which contributes to wife's marital happiness (Haavio-Mannila, 1986).   Gender integration of paid work thus increases cross-gender interaction both in the workplace and at home, which is particularly beneficial to women.   By working with men they become integrated into the higher status men's world of work.   This tends to increase their husband's participation in housework, which results in women reporting that they are happier in their marriage.

High social status has an encouraging effect on informal contacts between men and women--and also inside the gender groups--at work, even when one controls for gender-segregation in the workplace (Haavio-Mannila, 1987b).   The quality of work explains this result: higher status work is better paid, more autonomous and less routinized than lower status work (Kauppinen-Toropainen, Kandolin & Haavio-Mannila, 1988).   These characteristics of work probably provide good opportunities and also a need for social contacts both across and inside the two gender groups.

Exclusiveness of Marital Relationship

Marriage in many cultures expects spouses to limit their social and especially their erotic and sexual contacts to their marital partner.   However, there are many ways in which one can symbolically segregate men and women in society (Epstein, 1985).   One of the functions of these segregation mechanisms is to safeguard the exclusiveness of the marital relationship as the only legitimate cross-sex relationship permitted by society.

However, from the female's perspective, the integration of women into the society at large is functional, while the exclusiveness of the marital relationship is dysfunctional. In order to succeed at work or in politics, it is necessary for women to have contacts with men who still have more power and prestige in society than women. If the contacts of women are limited only to other women and to the husband, women are deprived of important social resources. For men, working in all-male settings results in a one-sided, harsh way of life in a homosocial macho culture which is reflected in interaction in the home. A heterosocial work environment, which fosters male and female interaction, would be more varied and human.

Classifications of Relationships and Roles

Social relationships at work and in the family can be classified into formal and informal ones. But formal and informal relations, as George Homans (1950) has demonstrated, are closely connected to each other. Homans suggests that joint activity in the formal work organization leads to informal interaction and sentiments of liking.

Male roles within the family can, according to Parsons and Bales (1956), be characterized as instrumental roles, while female roles are expressive. The instrumental-expressive distinction has been interpreted in terms of external vs. internal functions of the family system. Compared with Homans who saw different kinds of human behavior and feelings as being linked together, Parsons and Bales separated them into roles, which are performed by different persons.

METHODOLOGY

Research Questions

In order to assess the effects of cross-gender contacts in the workplace on the work and the family, several questions are posed:

* Do cross-gender relations at work vary according to family life-cycle? Do gender, parenthood and age influence social contacts between men and women outside the family at different stages of family life-cycle?

* How do spouses relate to each other throughout the family life-cycle?

⬛ What is the relative importance of cross-gender relations at work and in the family at different stages of the family life-cycle?

⬛ How satisfied and happy are husbands and wives at different family life-cycle stages?

Sample

The data were collected during 1986 by questionnaires mailed to members of eleven occupational groups living in southern Finland. The groups, selected on the basis of their sex ratio and social status, included the following occupations: dentists, journalists, architects, nurses for mentally handicapped children, mental health nurses, police officers, technicians, waiters, rubber and plastic workers, metal workers, and construction workers. Replies provided by 692 men and 667 women (married or cohabiting) respondents were analyzed. Data were collected from either the male or female in each family. In some cases, both members of the couple were respondents, but that was the result of occupational homogamy, i.e., both partners belonged to the same occupational group. In some groups (for example, women police officers) occupational homogamy is widespread. The data are not representative of the total population of Finland.

Family Life-Cycle Stages

The family life-cycle stages used in the analysis are those identified by Mattessich and Hill (1987) and Rexroat and Shehan, (1987):

⬛ Newly established couples (childless persons under 40 years of age)

⬛ Childbearing families (oldest child 0-6 years)

⬛ Families with schoolchildren and young adults (oldest child 7 years or more)

⬛ Families in the middle years or empty nest (children launched from parental household)

Sample Profile

The age structure of the occupational groups varied. There are a disproportionate number of journalists and technicians among women in the early family life-cycle stage, and many rubber, plastic, and metal workers in women's empty nest stage. Young, childless men are frequently mental health nurses and waiters, and architects are overrepresented in groups composed of men in the later stages of the family life-cycle. Males in the empty nest stage are over-represented in construction work.

Some of the occupational groups which are overrepresented in the early family life-cycle stages (e.g., female journalists and technicians and male mental health nurses and waiters) are less sex-typed than those belonging to later family life-cycle groups (male architects and construction workers). This is the result of younger people having chosen sexually non-traditional occupations more often than older people. The numerous women in the empty nest stage were industrial workers--traditional male jobs. These jobs will soon vanish due to automation; there is no recruiting of new young personnel into these jobs.

As a result of the age differences of the occupational groups, the gender structure of the workplace (Kauppinen-Toropainen, Haavio-Mannila, & Kandolin, 1984) of different life-cycle groups varies (Table 1). People in their early life-cycle stages are more often tokens, the only representatives of their gender group in the workplace, doing the same sort of work mostly with the opposite sex. Among men, the empty nest group includes the largest proportion of gender-segregated people, that is, persons who share work tasks and meet daily at work predominantly with those of the same sex.

Table 1. Percent of Gender Segregation of Work

Position In The Gender Structure	Young Childless Couples	Families With Oldest Child 0-6 Years	Families With Oldest Child 7 + Years	Empty Nest
Men	(N=101)	(N=149)	(N=345)	(N=62)
Segregated	19	20	19	32
Complementary	21	23	29	23
Integrated	40	38	41	40
Token	20	19	11	5
Women	(N=120)	(N=143)	(N=275)	(N=61)
Segregated	21	25	24	15
Complementary	18	20	21	28
Integrated	26	36	38	44
Token	35	19	17	13

Among women, those in the empty nest stage of the life-cycle
are more likely than women in other stages to have complementary
positions, i.e., they share work mostly with their own gender group
but are also in daily contact with members of the opposite sex.
Women in the later stages of life-cycle are often in integrated work,
i.e., they share work with approximately as many women and men. In
summary, newly established couples often work in token positions
together with members of the opposite sex. Among spouses in the
middle years, those in the empty nest stage are the most segregated
from members of the opposite sex in their work.

The social status of those in different stages of the life-
cycle varies across the 11 occupational groups (Table 2). Men in
the family establishment stage are seldom upper white collar workers
(dentists and architects). Half of them are lower white collar
workers, that is, nurses, police officers and technicians. Half of the
young childless women also work in these "pink collar" occupations
(Hower, 1977). Half of the women at the postparental stage of the
family life-cycle are manual workers, a proportion that is much
higher than found among women in groups composed of other stages.
Lack of gender segregation of work and pink collar status are thus
characteristic of men and women in newly established families and
families with small children. Husbands and wives, whose children
have already left home, often work in gender-segregated or complementary jobs. If they are women, half of them are industrial
workers. Respondents whose children are school age or older, but
still at home, have jobs in-between.

The variation of the sex structure and social status of the
family life-cycle groups restricts generalizations about the influence

Table 2. Percent of Families in Various Social Statuses

Social Status	Young Childless Couples %	Families With Oldest Child 0-6 Years %	Families With Oldest Child 7 + Years %	Empty Nest %
Men	(N=101)	(N=150)	(N=346)	(N=64)
Upper White Collar	26	43	41	45
Lower White Collar	51	41	41	33
Manual Worker	23	15	18	22
Women	(N=132)	(N=143)	(N=279)	(N=61)
Upper White Collar	37	46	44	33
Lower White Collar	52	42	34	18
Manual Worker	11	12	22	49

of the family life-cycle stage on cross-gender relationships at work and in the family. The real explaining factor may be, however, the gender structure of the workplace or the occupational status, rather than the stage of the life-cycle.

RESULTS

Gender Segregation in Work Relations

Informal social relations between men and women at work range from chatting, helping with personal problems, friendship, love, sex and marriage to sexual harassment. In this chapter, data analysis will be limited to cross-gender friendship, love and harassment. Other types of cross-gender relationships at work have been, to some extent, analyzed previously (Haavio-Mannila 1987a; 1987b).

As can be seen in Table 3, a greater number of husbands have cross-gender friends at work than do wives (55% vs 42%), in each stage of the family life-cycle. The gender gap is largest (22%) among parents of small children, reflecting young mothers' preoccupation with family life at this stage. However, for men, this stage in the family life-cycle is characterized by intensified friendly contacts with women at work. The result cannot be interpreted by referring to the gender structure of the workplace, which in this life-cycle group, is similar for men and women (Table 1). The different gender roles in childbearing families probably account for the large gender gap. In both gender groups, spouses in their middle years, and at the empty nest stage, have the least number of friends of the opposite sex. Cross-gender friendship may be looked upon as dangerous to the marital relationship by the older generation, which is probably quite traditional and bound by puritan sexual morality in its value orientation.

Being in love with or attracted to a co-worker, or someone met at work, does not vary much according to gender (11% of the husbands and 12% of the wives were in love at the time of the study) nor family life-cycle stage. Young childless wives reported the most workplace romances: 18% of them were in love with someone from work (See Table 3). About half of all respondents had experienced a workplace romance. More men than women reported having sex with the co-worker, and they also reported more extramarital romances (Haavio-Mannila, 1987a).

Using a 6-point scale (where 0=no cross-gender co-worker to 5 = co-worker loves me very much), husbands and wives in the early stages of the family life-cycle, as compared to those in the latter stages, reported that their closest cross-gender co-worker loved them at least a little (scores 2-5), and that they considered this co-worker to be important to them (scores 3-5 on a similar 6-point

Table 3. Percent of Cross-Gender Relations at Work

	Young Childless Couples	Families with Oldest Child 0-6 Years	Families with Oldest child 7 + Years	Empty Nest
Men	(N=101)	(N=149)	(N=224)	(N=63)
Women	(N=130)	(N=139)	(N=278)	(N=60)
Has cross-gender friends at work				
Men	51	59	56	46
Women	43	37	41	35
Is currently in love with a co-worker				
Men	11	14	10	11
Women	18	12	10	12
Is loved by one's best cross-gender co-worker and considers him/her to be a close and important person				
Men	20	16	13	8
Women	24	12	11	7
Has been sexually harassed at work during last 24 mo.				
Men	35	28	20	15
Women	78	70	26	17

scale). Thus while being in love with a co-worker is not related to the stage in the family life-cycle, being the object of co-worker's love and/or feeling close to this co-worker decreases as one moves towards the latter stages in the family life-cycle (See Table 3).

There is no gender gap in being loved by a co-worker of the opposite sex: 14% of both men and women reported being loved by a co-worker and considered him or her to be close and important. Five percent were objects of co-worker love without thinking that the lover was close or important to them.

Wives were more often objects of sexual harassment (30% had been harassed at work during the past two years) than husbands (24%). The gender gap was largest in the child bearing and rearing stages (See Table 1). Young childless people of both sexes were harassed most often, parents in their middle years least often. Here age probably explains the variation. Because young people are thought to be sexually attractive, they easily become the target of sexual advances, and sexual teasing.[1]

Gender Segregation in Family Relationships

At home, men and women still partly follow traditional division of labor and power according to lines schematized by Parsons and

Bales (1956). However, in Finland among the 11 occupational groups studied, only a minority of husbands devoted more time to paid work than their wives, and the proportion of families where husbands had more influence on family decisions than wives was even smaller. The patriarchal family is not the "norm" for family life.

Even though husbands seldom have more power in family decisions than wives, traditional family patterns still prevail in the division of daily domestic tasks. In a large majority of the families, the wife does more housework than the husband.

The influence of gender on the reports of cross-gender relations in the family is complicated by the fact that all women respondents are working for pay, whereas 14% of the wives of the male respondent are not employed. Likewise, while all the male respondents are employed, 8% of the husbands of the women in the sample are not economically active; they are students, unemployed, retired or just at home. Among wives in the empty nest stage, 26% have husbands who are not employed as compared with 21% of the males whose wives are not employed.

Husbands evaluate their work commitment to be a little stronger than wives do, perhaps because there are more "housespouses" in their families. A large gender gap emerges from conceptions about family power relations. In each life-cycle group, husbands evaluate their power in important family decisions to be greater than wives do. Among those in the empty nest stage, as many as 37% of the husbands but only 10% of the wives think that the husband has more to say than the wife in important decisions concerning the family (See Table 4). This gender gap may result from the large proportion of non-employed spouses in the families in their middle years (empty nest). This increases husbands' power in the family for the employed men studied here, and decreases it in families where the male is a "househusband."

While there has been a tendency for both spouses to overestimate their share of the housework (Berk & Shild, 1980; Haavio-Mannila, 1980), the perceptions of men and women regarding the division of domestic work in this study show a high level of agreement.

The stage in the family life-cycle influences the cross-gender relations within the family. Young, childless husbands equally share the housework with their wives. It is when the children arrive and the amount of domestic work increases, that traditional gender role patterns emerge. Husbands are not as willing as wives to increase their participation at home. Husbands' work devotion is, according to the replies given by wives, highest in families with children. Husbands in these families concentrate on paid work, and let their wives take care of the home.

It is also possible that gender equality among childless families might be reflecting the value of gender equality that characterizes

Table 4. Cross-Gender Relations in the Family, in Percent

	Young Childless Couples	Families with Oldest Child 0-6 Years	Families with Oldest child 7 + Years	Empty Nest
Men	(N=99)	(N=149)	(N=340)	(N=63)
Women	(N=130)	(N=142)	(N=275)	(N=58)
Husband more devoted to paid work than wife				
Men	35	43	40	40
Women	30	38	37	27
Husband has more influence on family decisions than wife				
Men	25	30	30	37
Women	15	15	19	10
Wife does more housework than husband				
Men	59	72	77	84
Women	62	71	81	72
Spouse loves the respondent very much				
Men	73	61	49	40
Women	78	70	61	60

this new generation. Further research on families in the childless stage will be necessary in order to assess whether it is non-traditional values, or the stage in the life-cycle that is influencing these new gender relationships.

In Table 4, the last aspect of family life, the feeling that one is very much loved by the spouse, is presented. There is a systematic decline from the earliest to the latest stage of the family life-cycle. Love in the marriage declines with age in a manner similar to being loved (and sexually harassed) by co-workers. It is possible that this reflects one's own perceptions of being loved and harassed. We must recognize that it is the respondents' perceptions of other peoples' feelings and doings, not the more objective measures of what the others actually feel and do, that is being measured.

Relative Meaning of Work and Family Relationships

In order to answer the research question about the relative importance of cross-gender relations at work and at home, a typology was constructed. It is based on answers to questions about the closeness and importance of one's relationship to the spouse and the closest cross-gender co-worker.[2] The distribution of husbands and wives according to their spouse co-worker orientation is presented in the following typology (See Figure 1).

	Close Integrated	Distant Spouse-centered
Close		
Men	34%	47%
Women	28%	56%

	Co-Worker Centered	Isolated
Distant		
Men	7%	12%
Women	5%	11%

Figure 1. Relationship to the Closest Cross-Gender Co-Worker

Men are more likely to have close social ties both in marriage and at work than are women. Women more often report feelings of closeness only to the spouse. There is a striking gender lag in being integrated with both the family and the work community in families with small children (See Table 5). Parenthood has a much greater limiting influence on the feelings of mothers than on fathers. Young childless husbands do not differ from young fathers in their "spouse-co-worker orientation," but there is a large difference between young childless wives and mothers of small children. The latter are much less integrated and more spouse-centered.

Table 5. Relative Importance of Spouse and Closest Cross-Sex Co-worker, in Percent

Spouse coworker Orientation	Young Childless Couples	Families With Oldest Child 0-6 Years	Families With Oldest Child 7 + Years	Empty Nest
Men	(N=101)	(N=150)	(N=346)	(N=64)
Integrated	40	41	31	23
Coworker-oriented	5	5	8	8
Spouse-oriented	46	49	46	58
Isolated	9	5	25	11
Women	(N=132)	(N=143)	(N=279)	(N=61)
Integrated	41	29	24	16
Coworker-oriented	3	2	6	7
Spouse-oriented	52	62	55	61
Isolated	4	7	15	16

Even when women work as many hours and earn as much as men (characteristic of childless and young-father families), motherhood results in a decline in cross-gender work relations. Having multiple close relationships at work and at home seems to be easier for fathers than for mothers. Isolation from close marital and work relationships increases as one moves towards the later stages in the family life-cycle, resulting in scarcer and less intense contacts.

Work and Family Satisfaction

Satisfaction with work. Satisfaction with work was measured with a sum-scale consisting of reactions to the following four statements: On the whole I am very satisfied with my job; I often think of leaving my job; I now work clearly in my own field; and I often feel great after work--I feel I have accomplished something. The five response alternatives range from 5 = agree strongly to 1 = disagree strongly. The reliability of the sum scale is .81.

There were no gender differences in work satisfaction. Neither were the differences between stages in the family life-cycle statistically significant (See Table 6). Gender and family roles do not seem to be important determinants of work satisfaction, which are mainly influenced by characteristics of work and the workplace. In contrast to the results above, "satisfaction with the division of housework in the family" was strongly influenced by both the gender and stage in the family life-cycle.

Housework satisfaction. Satisfaction with housework was measured by the question: "How satisfied are you with the division of housework in your family?" Responses ranged from 5 = very dissatisfied to 1=very satisfied. Women are significantly less satisfied than men with the division of housework in the family. Both husbands and wives are most dissatisfied during those stages of the family life-cycle when there are children at home. At that stage, the amount of housework is greatest (Rexroat & Shehan, 1987).

Traditional gender roles caused problems in the division of housework in families with children, with women doing most of the domestic work (See Table 4). This creates conflicts in wives' relations to paid work, and is expressed in their dissatisfaction with the division of housework in the family. Contrary to the situation in the United States (Rexroat & Shehan, 1987), most married women in Finland continue to work full time until the age of retirement (Statistical Surveys, 1984). In this sample, which consisted of employed wives and husbands, wives' work hours were approximately a half-an-hour less per day than husbands'. Furthermore, these hours did not vary by stage in the family life-cycle.

Wives do not reduce the hours worked outside the home in order to compensate for the uneven division of housework between the spouses during the child-rearing stage. Therefore, one solution to the conflict between family and work is the equal participation of both spouses in domestic work at home. This should be especially appealing since it increases wives' satisfaction with both the division of housework and the marriage (Haavio-Mannila, 1986). Unfortunately, this solution is practiced only in a minority of families.

Marital happiness. The marital happiness scale consisted of the following four items: "How do you regard your marriage or cohabitation? (a 7-point scale ranging from very unhappy to perfect); "Are you satisfied with your sex life with your spouse?" (a 5-point scale ranging from very unsatisfied to very satisfied); "Is your marriage or cohabitation based on sharing your entire life experiences and do you want to be together as much as possible?" (yes-no); and "Is your marriage or cohabitation a vital love relationship?" (yes - no).

The four items were reclassified so that each of them has the same weight on the total scale. The reliability of the sum-scale is .73. Husbands are, on the average, as happy as wives in their marriage. Marital happiness declines systematically from the earliest to the latest stage in the family life-cycle. Furthermore, life happiness and self-esteem decrease, when family life-cycle proceeds towards its later stages (See Table 6).

Table 6. Satisfaction and Happiness, Scale Means

	Young Childless Couples	Families With Oldest Child 0-6 Years	Families With Oldest Child 7 + Years	Empty Nest
Men	(N=100)	(N=147)	(N=339)	(N=62)
Women	(N=132)	(N=140)	(N=269)	(N=59)
Work Satisfaction				
Men				
Mean	15.2	15.4	15.3	15.8
Deviance	3.0	3.3	3.2	3.3
Women				
Mean	14.9	15.7	15.7	15.3
Deviance	3.5	3.1	2.9	3.9
F = 1.17 ns.				
Unsatisfaction with the division of housework in the family				
Men				
Mean	1.9	2.0	2.0	1.9
Deviance	0.9	0.8	0.8	0.8
Women				
Mean	2.2	2.6	2.6	2.3
Deviance	1.0	1.2	1.1	1.1
F = 11.41, p<.0001; 1 < 2,3*** Men < Women***				
Marital Happiness				
Men				
Mean	14.4	13.7	13.0	12.6
Deviance	3.1	2.9	3.1	3.0
Women				
Mean	14.3	13.6	12.9	12.2
Deviance	3.1	3.1	3.1	2.8
F = 5.03, p<.0001; 1 > 2,3,4, 2> 3,4*** Men vs. Women ns.				

The decline of marital and other kinds of happiness with aging may be related to the decrease of positive social contacts at work (See Tables 3 and 5); to the fading away of the romantic love in marriage (See Table 4); and to factors not discussed here, for example, death of family members and friends, decline in physical capacity and attractiveness, and illness.

CONCLUSIONS

Cross-gender relationships at work vary through the family life- cycle partly because younger people have more gender-integrated jobs. Thus at the earlier stages husbands and wives experience both more love and more sexual harassment by their cross-gender co-workers. Having cross-gender friends and workplace romances does not, however, vary much through the family life-cycle. The only exceptions to this stable pattern are, first, that at the postparental stage cross-gender friendship is rarer than during earlier family life-cycle stages, and second, that young childless wives are most likely to be in love with a co-worker.

The gender gap in heterosocial relations at work is deepest in families with children. It seems to be easier for fathers than for mothers to maintain close ties with opposite sex co-workers. At the family establishment and the empty nest stages, gender roles are not as differentiated as at the child-rearing stages.

The stage in the family life-cycle has only a minor effect on marital power and the division of housework. Conflicts over the division of housework are most severe in families with children. Love in marriage decreases when one moves from the earlier to the later stages of the family life-cycle. The weakening of emotional ties can also be seen in the increasing isolation from close and important ties with cross-sex co-workers and the spouse at later life stages. This detachment from social and emotional closeness may account for the reported low levels of marital satisfaction towards the end of the family life-cycle. Work satisfaction does not appear to be related to stage of the family life-cycle.

Sex at work has scarcely been studied in the sociology of work (Hearn & Parkin 1987). Erotic and sexual relations at work are not rare, even among married people. An earlier study found that 30% of married husbands and 22% of married wives had become attracted to, or fallen in love with a co-worker, and about half of these workplace romances evolved into a sexual relationship. The likelihood of falling in love at work does not diminish with age: there is very little connection between extramarital workplace romances at present and life-cycle stage. It is not possible to maintain only formal ties with co-workers, as informal interaction and sentiments of liking are unavoidable consequences of men's and women's joint activities in the formal organization of work. For the Finnish family, this is apparently not a threat as long as the couple considers the marital relationship to be a close and important aspect of their family life. ▓

ENDNOTES

1. In this study, we used similar items to measure sexual harassment in Finland as were used in the study on sexual harassment in U.S. federal workplaces, see Högbacka *et al.*, 1988.

2. The replies were given on a scale from 0 = not relevant to 5= very close and important.   The cutting point for the spousal relationship was 5 = close and 0-4 = distant.   For the co-worker a milder scale was used: scores 3-5 were classified as close and 0-2 as distant.

REFERENCES

Berk, S., & Shild, A.   Contributions to household labour: Comparing wives' and husbands' reports. In S.F. Berk (Ed.), *Women and household labor.*   London: Sage Yearbooks in Women's Policy Studies 5, 1980.

Epstein, C.F.   Ideal roles and real roles or the fallacy of the misplaced dichotomy.   In *Research in social stratification and mobility*, 4, Greenwich, CT: JAI Press Inc., 1985.

Haavio-Mannila, E. Kodinhoitotehtävien jakautuminen perheessä. Division of household tasks in the family. *Sosiologia*, 1980, 17(3), 185-194.

Haavio-Mannila, E. Influence of work place sex segregation on family life. Paper presented at the Conference on Changes of Family Patterns in Europe, Dubrovnik, October, 1986.

Haavio-Mannila, E. Cross-gender relations at work and in the family. Paper presented International Symposium on Nordic Intimate Couples, Hässelby Castle, Stockholm, 1987. (a)

Haavio-Mannila, E. Homo-och heterosocial interaktion på arbetsplatsen.   Working papers No. 40, Department of Sociology, University of Helsinki, 1987. (b)

Hearn, J., & Parkin, W. *Sex at work: The power and paradox of organization sexuality.* Kent: Wheatsheaf Books, 1987.

Hess, B., & Sussman, M.B. (Eds.), *Women and the Family: Two decades of change.* New York: Haworth Press, 1984.

Högbacka, R. *et al., Sexual harassment in Finland. Tasa-arvoasiain neuvottelukunnan monisteita.* Helsinki: Valtion   painatuskeskus, 1988.

Homans, G.C. *The human group.* New York: Routledge & Kegan Paul, 1950, (printed in 1951).

Howe, L.K. *Pink collar workers: Inside the world of women's work.* New York: Avon Books, 1977.

Kauppinen-Toropainen, K., Haavio-Mannila, E., & Kandolin, I. Women at work in Finland. In M.J. Davidson & C.L. Cooper (Eds.), *Working women: An international survey.* Chichester: John Wiley & Sons, 1984.

Kauppinen-Toropainen, K., Kandolin, I., & Haavio-Mannila, E. Sex segregation of work in Finland and the quality of women's work. *Journal of Organizational Behavior,* 1988, 9, 1, 15-27.

Losh-Hesselbart, S. Development of gender roles. In M.B. Sussman & S.K. Steinmetz (Eds.), *Handbook of marriage and the family.* New York: Plenum Press, 1987.

Mattessich, P., & Hill, R. Life cycle and family development. In M.B. Sussman & S.K. Steinmetz (Eds.), *Handbook of marriage and the family.* New York: Plenum Press, 1987.

Niemi, I., Kiiski, S., & Liikkanen, M. *Suomalaisten ajankäyttö.* (Use of time in Finland). Helsinki: Central Statistical Office of Finland. Studies No. 65, 1981.

Parsons, T., & Bales, R.F. *Family socialization and interaction process.* London: Routledge & Kegan Paul, 1956.

Pleck, J.H. The work-family role system. In P. Voydanoff (Ed.), *Work & family.* Palo Alto, CA: Mayfield Publishing Company, 1984.

Reskin, B.F., & Hartmann, H.J. *Women's work, men's work: Sex segregation on the job.* Washington, D.C.: National Academy Press, 1986.

Rexroat, C., & Shehan, C. The family life cycle and spouses' time in housework. *Journal of Marriage and the Family,* 1987, 49, 4, 737-750.

Sexual Harassment in the Federal Workplace. U.S. Merit Systems Protection Board. Washington, D.C.: U.S. Government Printing Office, 1981.

Sussman, M.B., & Shanas, E. Family and bureaucracy: Comparative analysis and problematics. In *E. Shanas & M.B. Sussman (Eds.), Family, bureaucracy, and the elderly.* Durham: Duke University Press, 1976.

Sussman, M.B. Family organizational linkages. In H. Harbin (Ed), *The psychiatric hospital and the family.* New York: Spectrum, Inc., 1983.

Statistical Surveys, Position of Women. Helsinki: Central Statistical Office of Finland, 1984.

▓ 16 ▓

HOUSEHOLDS OF THE ELDERS

John Mogey

INTRODUCTION

Research in social gerontology looks at aging as a set of social problems and as a consequence sees the aging process as a deviant rather than as an expected set of outcomes. Within this context, policies that administer scarce public resources such as medical skills or entitlements that support personal life styles, approach aging as something to be controlled or even diverted.

Social theorists also model the area as a set of structural problems; they deduce that bureaucratic service groups, in which responses follow on commands, can organize more efficiently for highly specific actions, and that primary groups can more efficiently meet diffuse and personal needs. Replacing the older term, social group, with the term network directs the attention of the reader towards observations of behavior rather than to the norms that underlie group interaction. Formal networks, as in hierarchal structures, contrast with informal networks and, using these concepts, the behavior of a community of elders can be seen to combine actions that belong to both network types. Independent living requires that informal and formal networks interact continuously to form the structure that Parsons calls the "societal community." This concept is the arena for the life styles of the elderly, retired from work, from child rearing, from school affairs, but still involved in networks that are both formal and informal (Cafferata, 1986; Litwak, 1985; Parsons, 1971; Sussman, 1981; 1982; Wenger, 1984).

213

Households

In all modern industrial societies the basic network of the societal community is the household. The norms underlying households are those of the nuclear family (Parsons & Bales, 1955; Sussman, 1959; 1979). Specifically, the norm for the formation of this sort of family is that those about to marry choose each other. In these societies, parents do not select spouses for their children, although they may try to influence their choices (Gillis, 1985; Goode, 1963; Hajnal, 1982; Macfarlane, 1986; Smith, 1981; Sussman & Burchinal, 1962). To reach the goal of free bargaining over marriage decisions between adult members of society, the necessary social skills should be nurtured between residents of each household. We expect the support of the individual, or the self, to continue after marriage in the new household and to be an essential aspect of the social structure of the nuclear family household. Using an earlier terminology, within households we should observe unbalanced reciprocity, a sort of altruism or sharing, sometimes referred to as amity (Farber, 1981; Fortes, 1969; Litwak & Selenyi, 1969; Sahlins, 1965; Sussman, 1977).

Many economists also see the household as a network that makes collective decisions about the distribution of its resources (Rosenzweig, 1986). Others, using exchange or bargaining theory, approach the household as an aggregate of persons where residence for each individual represents a separate decision about the marginal utility of membership for their own satisfaction level (Becker, 1973). Since households persist as long as the demands for biological, cognitive, and emotional support of members are met, and since sharing of the full resources of the network calls for commands and bargaining which are common to hierarchal and market structures, we can say what really distinguishes the nuclear family household from other sorts of primary groups in the societal community is the sharing of resources through the norm of amity.

Neighborhoods

Beyond the household, residence brings into play neighborhood interactions together with associations with others in interest groups such as churches and clubs, and with friendships with specific others. In that voting and taxes are based on residence as well as on citizenship, households link people to political structures. Residence also determines, for the most part, the locations where households purchase goods and services by bargaining in markets. Distinctions of social class, a major interest of social theory, do not apply within the household, though they do help to explain lifestyle differences between households.

Beyond the household, marriage and parenthood create wider kinship obligations. In these free choice marriage systems, the obligations felt by members of a kindred are roughly equal as between consanguinal and affinal kin. For social gerontology, kinship acts as a back-up to the resources of the household, since the norm of amity seems to apply to intergenerational and affinal relations, though it becomes less effective with distance from the face to face interactions of the household (Farber, 1981; Klatzky, 1972; Mogey, 1977; Sussman, 1979).

In this chapter, a national sample of elders will be used to demonstrate how households are influenced by kinship ties, and how interactions between households can be described as aspects of informal networks.

KINSHIP AND HOUSEHOLD FORMATION

From the national sample survey of 16,418 persons ages 55 and over, those over 65 were selected for this analysis (National Center for Health Statistics, 1984). As shown in Table 1, the selection produced a total of 11,319 persons. The composition of the households in which these elders reside form the basis for this analysis.

Two person households consisting of an elder and spouse constituted 44.5% of the total sample. Most of the single person households represented former married couple households, and together these two types of nuclear family households accounted for just over three quarters of all households of elders in the societal community.

Of those elders who lived with others, 1.1% lived with non-family or unrelated persons. All others lived in extended family households of some sort. The most frequent type of extended family households were those in which an elder lived with a child, or with a grandchild. These family of procreation households constituted 16.3% of the sample. No non-relatives lived in any family of procreation household nor in the other type of extended family household, the family of orientation household. Only 2.4% lived with a sibling, or less often, with a parent from the family of orientation of the elder.

A third type of extended family household, in which an elder lived with a collateral relative such as a niece, nephew, aunt or uncle, was even less frequent, representing less than 1% of the households.

When an elder lives in a household with a mixture of relatives from different types of kinship dimensions, some from the family of orientation, some from the family of procreation, and some collateral, this would be a fourth type of extended family household. Although this type of household had the widest chance to be formed, less than 3% of all households involved this sort of kin relationship. This is also the only household type that mixes kin with non-kin members (16% of these household contain a non-kin member), and are the largest households, averaging 4.1 as compared with 3.1 for the family of procreation households.

Table 1.   Percentage of Household Types with Elders by
            Mobility and Income

Household Type	Sample	At This Address[a] < 1 Yr.	Income $50,000 or More[b]
SAMPLE	(N=11,319)	(N=506)	(N=304)
NUCLEAR			
m. couple	44.5	1.5	56.6
single p.	32.0	1.4	4.0
EXTENDED			
procreat.	16.3	0.9	30.9
orientat.	2.4	0.1	0.9
collat.	0.7	0.1	0.0
mixed	2.7	0.2	7.6
NON-FAMILY	1.1	0.1	0.0

[a]Some 4.4% of the sample moved last year; 35% moved in the
past 10 years, so on average 3% to 4% of elders move house
each year.

[b]75.4% of single person households had incomes of $5,000 or
less in the past year, and so had 7.5% of family of procreation households; the high income households appear in the
table.

Source for all Tables: National Center for Health Statistics,
Supplement on Aging to the National Health Interview Survey,
1984. Public Use Tape No. 803076.

     We conclude that household formation among the elderly rests
on social ties that follow the form of amity or on kinship, since
98.9% of all persons 65 or older lived in such households.
     The impact of marital status on income is illustrated in Table
1. Of the 2.7% of households with the highest incomes, more than
half were married couple households, another 30% were family of
procreation households, and an additional 7.6% were mixed households. Among the 1,164 households (10.3% of the sample) with
annual incomes of $5,000 or less, 75% were single person households
and 3% were non-family households. This shows that poverty is
most common among those living alone, or with strangers.

The 1981 Census recorded that of the population 65 or older, 6% lived in group quarters (homes for the aged, nursing homes, hotels, barracks, etc.) and of those 85 or older, 25% lived in group quarters (U.S. Senate, 1985). This sample and census data confirm that about 94% of all elders live in family households.

Given that the normative preferred household type is the nuclear family one, then other household types represent a hierarchy of preferences, moving from extended family households to living with non-relatives either in a household, or in group quarters. Consequently, as individuals age, they follow a progression among household types. The rate of change is between 3% and 4% per year for those aged 65 and over. The decision to live in an extended family household, or with non-relatives only, presumably as a boarder, increases markedly for those over age 85 and for racial minorities (See Table 2). However, female elders tend to live alone, rather than change houses. Changes occur because of increasing frailty, illness, death of a spouse, or occasionally because of poverty: for example, in 80% of all single person households, the persons were female. Of 225 elders in the sample widowed for less than one year, 78% lived in single person households and 22% moved into family of procreation households. For the 2,345 elders widowed from 1 to 10 years, 67% lived alone, and 23% lived in family of procreation households.

Table 2.   Percentage of Household Types with Elders by Age, Race, and Sex

Household Type	65 and +	85 and +	65/+ and non-white	65/+ and female
SAMPLE	(N=11,319)	(N=809)	(N=972)	(N=6,745)
NUCLEAR	76.5	62.2	58.3	75.9
EXTENDED	22.4	35.7	39.1	22.8
NON-FAMILY	1.1	2.1	2.6	1.3

Among non-whites, the proportions living in extended family households, or with non-relatives, are much higher than for the sample as a whole. Among these less preferred household arrangements, non-whites live in households with collateral relatives, and with non-family people, about ten times as often as the rest of the population.

Supports For Elders in Households

Independent living assumes that persons can engage in interaction with others. To be an adult in an industrial society calls for interaction in a minimum of three separate social structures:

* Obedience to the rules of political authorities who operate command structures;

* Making purchases in markets of all sorts, that is, bargaining with others in exchange structures;

* Responding to the needs of others in kinship structures as in households and intergenerational reciprocity.

Since household composition has been shown to be 99% based on kinship, we may assume that helping between others in the same household rests on obligations of amity. Of course, members do bargain with each other and do give commands, but unbalanced reciprocity seems to be more significant. Of the 917 cases in 11,319 households, where an elder had daily help with getting into and out of bed, 11 had paid help with this activity of daily living (ADL): 9 of these were in Nuclear family households and 2 in mixed households. This represents the number of households with live-in servants, who give assistance to 1.2% of those in need or to .01% of the sample households. The obligation to help within these households rests on kinship.

Supports for Elders between Households

In informal networks, any member is expected to be capable of doing all the common activities necessary for everyday living. The fifteen specific tasks asked about in the survey are divided from this presentation into two classes:

1. Activities of Daily Living (ADL): bathing, dressing, eating, getting into or out of bed, walking around inside the house, walking outside the house, using the toilet, preparing meals in the house.

These tasks prepare the individual for social interaction; 60% to 70% of helping with them comes from relatives living in the house, most of it from the spouse; another 15% to 20% comes from relatives in other households. For purposes of analysis, the items "getting into or out of bed," and "preparing meals" were selected to represent the ADL indicators (See Tables 3 & 4).

2. Instrumental Activities of Daily Living (IADL): shopping, managing money, using the telephone, heavy housework, and light housework.

These tasks involve the actor with non-household persons; 55% to 75% of helping with them comes from relatives in the house, and another 25% from relatives living in other houses. Help with these IADL tasks is more often purchased in a market than for ADL tasks. Purchases of services for these tasks range from 25% for "heavy housework" to 2% with "managing money." Such paid helpers are used mostly by elders living in non-family households. For this chapter, "telephone" and "heavy housework" were selected as typical of the IADL tasks (See Tables 3 & 5).

Table 3. Number of Households with Elders That Reported Needing Help

Household Types	ADL Needs[a]		IADL Needs[b]	
	Bed	Meals	Phone	House
Nuclear				
Married couple	284	234	190	892
Single	314	207	89	1014
Extended Family				
of procreation	228	250	182	558
Non-family	20	26	19	50
Total in number in need	917	797	533	2703
Percent of sample	7.9	6.9	4.6	23.5
Percent of persons in need unable to do tasks	15.8	5.4	41.7	64.3

[a]Bed=Getting into or out of bed; Meals=Cooking meals in the home.
[b]Phone= Using the telephone; House= Doing heavy housework.

It might be supposed that elders who live alone would receive more help than elders in households with several people in them. By presenting the actual members of households in the sample in Tables 3, 4, and 5, we show that single person households report ADL needs at about the same frequency as married couple households and family of procreation households; all of them report that 30% of households have difficulty with the various tasks. For the two IADL tasks, single person households report less difficulty with telephoning (17% as compared with 34% or 35%): however, they more frequently report difficulties with housework (38% as compared with 20%) in family of procreation households, where there are younger residents. Furthermore, elders living in non-family households report difficulties in only 2% or 3% of households (See Table 3).

In fact, more single person households are helped by relatives than are other household types. Elders in non-family households are isolated from kinship help in all these tasks, and since they are poorer, even from paid help. In all the households combined, of those saying they needed help with any task, the percentage actually helped by another person varied from 24% (walking inside the house) and 55% (walking outside) to 91% (management of money). For management of money, only 2% of help came from non-relatives while 89% was help provided by relatives (See Tables 4 & 5).

Table 4. Sources of Help with ADL For Households With Elders

	Bed			Meals		
Household Type	Relative[a] Close	Relative Other	Non-Relative[b] Paid	Relative Close	Relative Other	Non-Relative Paid
Nuclear[c]						
m. couple	5	2	18	8	7	33
single	6	10	14	23	29	59
Extended Family of Procreation	7	4	19	5	6	26
Non-Family	0	0	2	0	2	2
Supports From All Sources, Total Households Helped.		326			118	
Of Those in Need, % Helped		35.6			14.8	

[a]Close Relatives = Spouse, Parent, Child only.
[b]There are in supports from all sources a few examples of help from unpaid non-relatives.
[c]Total Households = 11,319.

More descriptively, if being unable to get into and out of bed results in being bedridden, then 15.8% of the sample elders are bedridden and half of these live with married children in family of procreation households. If being unable to telephone means that the person is deaf, then 14% of those living alone are deaf, and 3% of those in non-family households are deaf (See Table 5).

Table 5.   Sources of Help with IADL For Households
With Elders

Household Type	Phone			Housework		
	Relative Close	Relative Other	Not Relative Paid	Relative Close	Relative Other	Not Relative Paid
Nuclear						
m. couple	1	0	7	35	85	170
single	8	13	7	91	197	342
Extended Family						
of Procreation	2	4	11	17	25	46
Non-Family	0	0	0	0	3	10
Total Support		333			2,192	
Of Those In Need, % Helped		62.5			81.1	

CONCLUSION

The societal community of elders depends for its continuity on three types of social interaction:

▓ Those underlying household formation and continuation;

▓ Those underlying the informal networks that give assistance between households;

▓ Those underlying formal networks that provide resources for life style continuity.

In this paper we show descriptively that households are formed around the normative obligations of the nuclear family and the kinship system. Support for elders within their household is almost entirely given by relatives of the elder. For "cooking at home," total amount of help from non-relatives in all 11,319 households was limited to 33 cases; 24 of these 33 cases were in non-family households. For this daily chore, then, 99.7% of all assistance to elders came from relatives. Help also came into the household from relatives living in other houses. Finally, about 20% of these outside helpers were non-relatives, and most of them were paid. Between households, IADL or social care tasks followed a similar pattern, with about 20% of social care provided by non-relatives.

If unpaid non-relatives were friends or neighbors, their maximum contribution was in the form of helping the elder with walking outside the home, a companion role: 8% of the 1,255 households experiencing this difficulty had help from friends or neighbors. If paid, nonrelatives bringing help into the house represented purchases of personal services in a market, then their maximum contribution was either 24% help with eating meals, or 25% help with heavy housework.   These data underscore the predominance of kin networks as a source of service to elders who were experiencing difficulties in managing their daily activities.   The only alternative available to most was the purchase of services; between 10% and 25% of households needed help with purchasing these services.   The interesting point made by all these data is that help given by other informal networks, such as personal friends, neighbors, associates in interest groups, while crucial in individual cases, was so small as to be ignored as a general rule.

The principal theoretical insights to be drawn from these data is that more explorations of the effects of kinship in industrial societies have to be made.   This paper augments a body of knowledge well documented in the literature over the past thirty years. In its detail it shows that kinship interaction differs from other interactions in informal structures.   Although based on unbalanced reciprocity, this norm has different consequences in households and between related households than it has in bureaucratic structures, where there is also unbalanced reciprocity.   Social theorists interested in the family need to add to their current statements about social class, authority, and exchange, a new set of theory statements about amity.   I suspect that these statements will integrate the concept of the self, and its care and support through informal interaction networks, with these other theories. ▨

In preparing this paper so much help and support came from Bernard Farber and Ione DeOllas, Arizona State University, and from Robert A. Lewis, Purdue University, that they deserve to be co-authors. The paper arose from a joint project with the Institute of Sociology, Hungarian Academy of Sciences, on kinship in two differing societies; the project is part of the cross-national program of the International Research and Exchanges Board, Princeton, N.J.

REFERENCES

Becker, G.S.   A theory of marriage, (Part I)   *Journal of Political Economy*, 1973, 81(4), 813-846.

Becker, G.S.   A theory of marriage, (Part II)   *Journal of Political Economy*, 1973, 82(2), 511-526.

Cafferata, G.L.   Caregivers of the frail elderly: A national profile. Paper presented at the American Society on Aging, 1986.

Farber, B.   *Conceptions of kinship.*   New York: Elsevier, 1981.

Fortes, M.   *Kinship and the social order.*   Chicago: Aldine, 1969.

Gillis, J.R. *For better, for worse: British marriages 1600 to present.* New York: Oxford, 1985.

Goode, W.J. *World revolution and family patterns.* New York: Free Press, 1963.

Hajnal, J. Two kinds of pre-industrial household formation systems. *Population and Development Review*, 1982, 8, 449-494.

Klatzky, S.R. Patterns of contact with relatives. Washington D.C.: American Sociological Association, 1972.

Litwak, E., & Selenyi, I. Primary group structures and functions: Kin, neighbors and friends. *American Sociological Review*, 1969, 34, 465-481.

Litwak, E. *Helping the elderly: The complimentary roles of informal networks and formal systems.* New York: Guilford Press, 1985.

Macfarlane, A. *Marriage and love in England, 1300-1840.* Oxford: Basil Blackwood, 1986.

Mogey, J. Content of relations with relatives. In J. Cuisenier (Ed.), *The family life cycle in European societies.* Paris: Mouton, 1977.

National Center for Health Statistics, National Health Interview Survey, 1984 Supplement on Aging, Public Use Tape #803076.

Parsons, T. *The system of modern societies.* Englewood Cliffs, N.J.: Prentice-Hall, 1971.

Parsons, T., & Bales, R.F. *Family, socialization and interaction process.* New York: Free Press of Glencoe, Inc., 1955.

Rosenzweig, M.R. Program interventions, intrahousehold distributions, and the welfare of individuals: Modeling household behavior. *World Development*, 1986, 14(2), 233-243.

Sahlins, M.D. On the sociology of primitive exchange. In M. Gluckman & F. Eggan (Eds.), *The relevance of models for social anthropology.* New York: Barnes and Noble, 1965.

Shanas, E., & Sussman, M.B. (Eds.), *Family, bureaucracy, and the elderly.* Durham, N.C.: Duke University Press, 1977.

Smith, R.M. Fertility, economy, and household formation in England over three centuries. *Population and Development Review*, 1981, 7, 595-622.

Sussman, M.B. Social and economic supports and family economists for the elderly. 1979, AoA Grant #90-A-316.

Sussman, M.B. The isolated nuclear family: Fact or fiction. *Social Problems*, 1959, 6(Spring), 33-340.

Sussman, M.B. Role of family and kin in social intervention. National Institute on Aging, 1981.

Sussman, M.B. National Research Planning Panel. Informal Supports Networks. Conference Proceedings, New York State Advisory Council, 1982.

Sussman, M.B., & Burchinal, L.G. Kin family network: Unheralded structure in current conceptualizations of family functioning. *Marriage and Family Living*, 1962, 24, 231-240.

U.S. Department of Commerce, Bureau of the Census, Current Population Reports, Series P-20, No. 371, Household and Family Characteristics, 1981.

U.S. Senate, Special Committee on Aging. How older Americans live: An Analysis of Census data. Sen. Report. 99-91, 1985.

Vital & Health Statistics, (No. 115), Kovar, M.G.,   Aging in the eighties.   (No. 116), Kovar, M. G., Age 65 and over and living alone; Contacts with family, friends and neighbors.   (No. 124), Stone, R., Age 65 and over: Use of community services. (No. 125), Havilak, R.J., Impaired senses for sound and light in persons aged 65 and over.

Wenger, G.C. *The supportive network: Coping with old age.* London: Allen and Unwin, 1984.

❋ 17 ❋

PROFESSIONAL CLIENT RELATIONSHIPS AND THE OLDER PATIENT

Marie R. Haug

INTRODUCTION

Professional-client relationships and the older patient, when viewed from an interdisciplinary perspective, form a complex cobweb of interactions that are affected by provider occupational and personal characteristics linked to patient health and personal characteristics. Despite the flood of social science literature on provider-patient relationships, there is a dearth of material on elderly patients, and most of what there is focuses on problems of compliance with physicians' recommendations among patients of all ages. Yet older people deal with many professionals other than physicians in the health care arena, and there are several issues other than following doctor's orders that need to be addressed.

To begin, consider the occupations that might be involved. In alphabetical order they are the dentist, nurse, pharmacist, physical therapist, physician and social worker. With which of these would one think the elderly interact the most? There are no data, but in light of the multiple medications prescribed for many of advanced years, not to mention over-the-counter drugs, a good guess would be a pharmacist. Perhaps next is a nurse, who is involved in almost all visits to a doctor's office, but who also has many more contacts than the physician in the hospital or nursing home, not to mention relationships as a private duty or visiting nurse.

Physicians undoubtedly come next, but note that except for times when the patient is in a hospital, when visits are usually daily, or in a nursing home when they may be monthly, the average contact is about six times a year (Verbrugge, 1983). Social workers may come next, followed perhaps by physical therapists, with dentists likely to be the providers with least elderly patient contact. In one study only 10% of those 65 and over had seen a dentist in the past year (Kandelman & Lepage, 1982).

DEMOGRAPHIC CHARACTERISTICS

Complicating matters further is the fact that each of the holders of these occupational roles may differ in demographic characteristics. They can represent varying age groups, ranging from 25 to 70 and older. They can be male or female, Black, white or Hispanic. Although they are now middle or upper-middle class, some of their values reflect the working, middle or upper class status of their parents. Similarly their religious backgrounds represent Catholic, Jewish, Protestant, or none.

Virtually no studies of professional-client relationships take these factors into account. Providers are seen as monolithic, unaffected by these demographic characteristics, which are supposedly outweighed by their professional standards. Such a situation seems highly unlikely, but again substantiating data are missing.

Clients vary too. The health problems that older people bring to a provider are by no means uniform. Primary consideration is the client's illness status. The complaint might be chronic or acute, or if the patient is asymptomatic, he or she might be coming for an annual physical. Without regard to clinical diagnosis, the patient's self assessed health, from poor to excellent, is another relevant factor. Similarly, the functional status of the patient is important, both in terms of activities of daily living (ADL) such as being able to eat, dress and toilet without help, and instrumental activities of daily living (IADL), such as using the phone and shopping. Again these can range from poor to excellent, with implications for interaction with providers.

Although patients' demographic characteristics match those of providers, their categories differ to a degree. Now, at the upper ranges of life, age cohort is significant. Those aged 65 to 74, the young-old, will differ from those 75 to 84, and from the old-old, those 85 and over. Gender, ethnicity (Black, white or Hispanic) and religion (Catholic, Jewish, Protestant or none), are similar factors among both providers and patients, but social class now applies to the current status of the individual, rather than to that of his or her family. Patients can consider themselves to be working, middle or upper socio-economic status.

The potential combinations of these characteristics when patient meets provider are mind-boggling. As a diversion into a mathematical exercise, the number of such relationships was calculated. Since each occupation could have a total of 288 possible joint demographic characteristics, and there are six occupations, there are 1728 variations possible on the provider side.

Among the clients, there are 48 different combinations of health statuses, and 216 varying demographic combinations, so that in all there are 10,368 different characteristic mixes. Relating provider and client variations produces 17,915,904 different professional-patient combinations.

Obviously some combinations would be rare, or non-existent. For example, how many Black male nurses, who are over 70 and come from middle class families of the Jewish faith, are there? Or how many Hispanic women over 85 without any functional deficits but in poor health, are upper class and Protestant? And how likely are these two to come together in a provider-patient encounter? Even leaving aside these real life anomalies, there are enough mixes to discuss to take the rest of my life time as well as that of all readers. Therefore, some limitations are in order. Based chiefly on the availability of data on dealing with the elderly patient, the balance of this paper will perforce focus on three provider occupations: dentists, nurses, and physicians. There is more information available on the latter than the former two.

Even with this focus, further limitations are necessary. Taking the demographics of just one occupation and one elderly client, a veritable spider web of relations is possible. Discussing all these inter-relationships is obviously out of the question. Accordingly, most of what follows will stress age and gender matches or mismatches between provider and patient, with some few references to ethnic status and social class, but only speculation on the effect of religious differences, on which at this point there are virtually no published references. The locale of an interaction is also worth brief consideration.

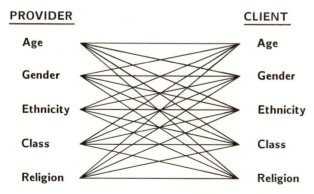

Figure 1. Provider/Client Relationships

There is one final complexity that refers to the context or quality of any interaction, what might be called the five interaction themes.

Interaction Theme Characteristics

- Congruence-incongruence
- Superordinate-egalitarian
- Technical-affective
- Dyad-triad
- Respectful-disparaging

Demographic congruence. Whether the demographic characteristics of the actors are congruent or incongruent is the first issue. Second, interaction can involve a superordinate-subordinate role set, with the provider as the superordinate actor. Or reflecting "the revolt of the client" (Haug & Sussman, 1968), it could be egalitarian, with both parties acting as equal partners. Third, the approach of the provider may be purely technical, or lean toward a more affective stance. Fourth, the relationship may involve a dyad, with only the provider and patient present, or a triad, with a family member accompanying the patient, as is often the case with the old-old. Finally the attitude of the provider toward the patient can be respectful or disparaging, with the latter arising from a provider's ageist biases. Such attitude variations on the part of the patient are undoubtedly rare: most of the elderly have been socialized to respect professionals, particularly physicians.

AGE FACTORS

Age is a provider characteristic that impinges on relationships with a patient. Among physicians most likely to deal with elderly patients (general practitioners and internists), 30% and 17% respectively are 65 and over, with 20% and 35% of their patients also elderly (Robert Wood Johnson Foundation, 1981). These older medical practitioners are unlikely to have been trained in geriatrics, so that their knowledge of aging processes is apt to be informal or experiential.

Unless they have kept up with the growing literature on the special health problems of aging, their knowledge is also likely to be outdated. They are almost certainly untrained in the social and behavioral sciences, which are relative newcomers to medical school curricula. Many health problems of the elderly involve their social contexts and require physician sensitivity to, and understanding of, the implications of such conditions if care is to be successful (Robert Wood Johnson Foundation, 1981; Eisenberg, 1985). Given these considerations, older physicians' competence in treating the old can be less than optimal, leading to possible misdiagnoses and treatment errors.

It has been pointed out (Stoeckle, 1987), that as the average age of the population increases, the average physician's age decreases. The sizable increase in recent years in the production of new physicians means there are now many younger doctors who of necessity will be seeing growing numbers of older patients. Although some of these younger practitioners may have been exposed to geriatric training, the generational disparity between provider and patient can very well cause difficulties in communication and mutual understanding. Younger physicians may find it impossible to realize the full effects of aged patients' impaired hearing, vision, and mobility on interaction with them. As a consequence, they can become impatient with securing a reasonable history or description of symptoms, or with the extra time needed for examination and explanation of treatment plans (Pendleton, Brouwer, & Jaspars, 1983).

Age distributions of dentists do not differ materially from those of physicians: one study estimated 14% over age 60 (Kiyak, Milgrom, Ratener & Conrad, 1982). Most had no exposure to geriatric dentistry, which even recently was taught formally in only 12% of dental schools (Ettinger, Beck, & Jakobsen, 1981). Problems of relationships to older patients may mimic those of physicians. However, they do not see the elderly as frequently as physicians do, on average only once a year (Verbrugge, 1983), either because such service is not covered by insurance or because many elderly are edentulous. If their false teeth are uncomfortable, they are likely to use over-the-counter preparations rather than visit the dentist. Moreover, elderly are rarely seen by dentists in hospitals or nursing homes since dentistry as a rule requires sophisticated, non-transportable, equipment. (Claus, 1982).

Unlike doctors and dentists, some of whom continue their practices well into old age--even after reaching 70--nurses tend to leave practice earlier. Sometimes they leave for a period to raise their children and then return in their middle years. Younger nurses, like younger doctors, will meet many elderly patients in physician offices and hospitals, while older nurses usually find work chiefly in nursing homes. Gerontological nursing appears to have been more commonly taught in schools of nursing than geriatrics in schools of medicine.

As for patients, it is important to recognize that there is perhaps more heterogeneity among the conventionally defined elderly group, those 65 and over, than among younger persons. At least three somewhat internally similar groupings, however, can be identified, representing three major birth cohorts, with different characteristics and historical experiences (Bloom & Speedling, 1981; Riley, 1985).

Reliable information on the current old-old, those now 85 and over, is just beginning to become available. They suffer more economic losses, have a lower educational level, suffer more cognitive impairment, and not surprisingly, have a lower physical health and functional level than younger cohorts (Haug & Folmar, 1986). They are heavier users of health care, both ambulatory and in such institutions as hospitals and nursing homes (Verbrugge, 1983).

Moreover, their health care experiences differ from those of younger cohorts because they were born before the cascading scientific discoveries dramatically changed the practice of medicine in mid-century. They went through two major economic depressions. Some, particularly peasant immigrants and rural Blacks, may have had virtually no experience with doctors or hospitals in their youth, except when their own elderly relatives were dying. Lack of physician contact could make them most fearful and suspicious of interactions with such practitioners, while their illnesses, pains, and other troubles make them particularly anxious to secure health care.

Younger cohorts, those born in the 1920s often have more education and exposure to new sophisticated treatment potentials, as well as more experience with physicians in their younger days. However, they still might be skeptical of the efficacy of medicine and physician care (Kleinman & Clemente, 1976).

A concomitant of clients' age is their attitudes toward the health effects of aging. The attribution of ailments to age will result in failure to consult a physician for treatable conditions (Kart, 1981). Also the elderly consider it "normal" to have dental problems. They think it useless to seek help (MacEntee, 1984), with one consequence that the average number of healthy teeth found in one sample of older persons was only 2.2, and 72% were entirely toothless (Kandelman & Lepage, 1982).

Age and professional-patient relationships. The impact of age on the course of the relationship can be assessed in terms of the five interaction themes noted earlier. Age characteristics congruence is rare as the data on age distributions of doctors and dentists demonstrates. Older doctors are more likely than younger ones to adopt a superordinate role (Haug & Lavin, 1981), rather than an egalitarian one in which patient and professional negotiate a mutually agreeable course of action (Katon & Kleinman, 1981).

Recent medical school graduates tend to espouse this more egalitarian style (Lavin, Haug, Belgrave, & Breslau, 1987), and young physicians have been known to dress informally, in jeans and sneakers, to cut the ambience of authority surrounding more formal attire (Stoeckle, 1987; Shorter, 1985). On the other hand, an authority-based stance may be more comfortable for the elderly patient accustomed to physician dominance and less happy with participation in decision making (Haug, 1979). But firm data are missing. One can only speculate that these relationship preferences can be found with respect both to physicians and dentists.

As already reported above, age congruence between aged patients and nurses is highly unlikely, given age distributions in the nursing profession. There are some indications that nurses, because of their responsibility for educating patients (Winslow, 1976), have assumed a superordinate role in the past, which, according to one study, is now changing. Prior "authoritarian lines" in which hospital regulations were rigidly applied is giving way to a "mutual learning experience" (Yuen, 1986; Evangela, 1968), in which decision making is shared with the patient.

Patient response to this change by age is not noted, although it is significant that recent articles on this change appeared in a symposium on patient compliance, suggesting the goal was to persuade the patient to follow recommendations of health personnel. Nurses, like doctors (Waitzkin & Stoeckle, 1976) and dentists, will find it hard to share information with patients, because it diminishes the power gained from their expert knowledge and, thus, their authority and influence (Winslow, 1976). The importance of teaching the elderly is one of the few references to patient characteristics in connection with the nurses' educational role (Dall & Gresham, 1982).

Explicit discussions of technical versus affective approaches by age, the third theme, are not found in the literature. One study reported more courtesy accorded by physicians to the elderly (Hooper, Comstock, Goodwin, & Goodwin, 1982), but another reported less positive rapport (Stewart, 1983). Furthermore, physicians have been found to spend less time with patients 65 and over (Keeler, Solomon, Beck, Mendenhall, & Kane, 1982). Since an affective approach, which requires concern about social situations and feelings, can take more time than the strictly technical, the finding implies that an affective stance is rare. Data on dentists are missing. Nurses on the other hand, have been advised to be accepting, sensitive and alert to patient feelings (Schoen, 1967), all of which imply affective rather than more technical styles.

The problems in professional-client relationships generated by the inclusion of a third party, such as a family member, in the situation has been perceptively discussed by Rosow (1982) and Coe and Prendergast (1985). Their studies refer only to doctors, undifferentiated by age, and show the danger of a very old patient being ignored as doctor and family member form a therapeutic coalition. Data on dentists do not exist, but the same issues could arise. Nurses, on the other hand, deal with patient families all the time and are urged to involve family members as fully as possible in meeting the needs of the aged ill (Evangela, 1968). Whether or not involving the family in the plan of care tends to isolate the patients and contribute to untherapeutic dependency is not discussed.

Finally, and perhaps the most important relationship issue, is whether or not the professional is respectful to the old client, or disparaging--either explicitly or implicitly. Here is where ageism comes into play. The elderly, already diminished in their well-being by physical, emotional, and functional problems, have a major need for respect from professionals, and a feeling that their dignity is recognized (Marshall, 1981). It should be pointed out that all three providers have been found to hold negative views of the elderly and to dislike treating them. Prejudice against and denigration of the aged still appear to be widespread among physicians (Ford & Sbordone, 1980) and other health professionals (Kosberg, 1983).

Reasons for physician ageism include the prevalence of chronic illness among the elderly, which goes against the cure orientation of professional training, and because of likely exposure to death and dying, accompanied by unresolved personal fears about their own demise (Butler, 1975). Perjorative terms such as "crock" and "gomer" are common among medical students as applied to the old and feeble with multiple ailments. These are attitudes that can carry over into later practice. Patronizing physician comments, such as "what can you expect at your age" (Butler, 1975) or the "bogus, unearned familiarity" of using first names in addressing patients (Conant, 1983), diminish the old person's self esteem and sense of worth.

A clue with respect to ageism among dentists is found in one study that showed their ideal patient was under 55 (Collett, 1969). Negative stereotypes of the elderly are found among dental students as well as those in practice (MacEntee, 1984; Claus, 1982). Nurses appear sensitive to the issue of patient dignity, noting that treating old people like children is degrading, and that respect is an essential factor in the plan of care (Moses, 1967). Because of the invasive and personally embarrassing procedures necessary in much hospital nursing work, it is particularly important that they be carried out with sensitive attention to maintaining patient dignity and self esteem (Green, 1986). Yet, as noted above, nurses are by no means immune to ageist prejudices and negative stereotypes about the elderly (Ciliberto, Levin, & Arluke, 1981), and are unwilling, as reported in several studies, to work with geriatric patients (Ingham & Fielding, 1985).

The Gender Issue

Over 90% of physicians are male, while nearly 2/3 of their patients are female (Stoeckle, 1987). The percentage of women dentists is also very low, although here, as in medicine, women are entering the field in increasing numbers. Nearly a quarter of a recent graduating class in a major medical school were female (Lavin et al., 1987), but currently only 8% of practicing physicians are female (Stoeckle, 1987). Even women already in practice are likely to be in specialties like pediatrics or psychiatry with few or no elderly patients, rather than in family or internal medicine where old people are more often seen (Weisman, Levine, Steinwachs, & Chase, 1980). This specialty mix appears to be changing however. The career plans of male and female medical students are beginning to converge (Shapiro & Jones, 1979; Bluestone, 1978).

Unlike medicine and dentistry, nursing is overwhelmingly female, although there are beginning to be a few male nurses. The cost of nursing, that of being seen as a female occupation, has been a lower public status. In other words, it is an apprenticeship relationship with the male doctor and limited work autonomy (Church & Poirier, 1986).

Elderly patients, reflecting the general elderly population, are predominantly female. The demographics demonstrate that for every 100 women age 65 and over there are only 67 men, and at age 85 and over, the fastest growing segment of the population, there are only 42 men for every 100 women (US Senate Special Committee, 1985). Women are more likely than men to suffer from chronic conditions, like arthritis and diabetes that are not immediately life-threatening, while men more commonly are subject to fatal heart attacks and strokes. These differences in illness patterns contribute to the greater longevity of women, but they also may account for their higher rates of physician utilization (Verbrugge, 1982). However at advanced ages, 75 and over, there is evidence that women make about the same number of physician visits per capita as men, 6.4 per year, while their hospital stays are longer (Verbrugge, 1983).

Gender and professional-patient relationships.   The effect of the gender factor on provider-client characteristic congruence is clear. Except for nurses, incongruence is the rule. Women patients see male physicians and dentists. Only in nursing are they cared for by women. It should be recognized, however, that elderly women, who have a habit pattern of being treated by male physicians, may actually prefer that situation. For them, being attended by a woman doctor or a male nurse could be stressful.

Although there are no data to support this speculation, there is evidence that the revolt against treatment by male gynecologists came chiefly from younger women (Boston Women's Health Book Collective, 1973). Gender match between patients and providers has been found to have a positive effect on clinical interactions since it fosters rapport and disclosure (Weisman & Teitelbaum, 1985; Levinson, McCollum, & Kutner, 1984).

In at least two midwest studies (Heins, 1979; Lavin, 1982), women physicians appeared to prefer a more egalitarian approach to patients than did males. If license to interrupt another person's discourse is a symbol of superordinate position, then women physicians were found to be less dominating in one study that showed they were far less likely than male physicians to interrupt patients as they were trying to explain their problems (West, 1984).

Undoubtedly, such interjections can interfere with the practitioners' thorough understanding of the relevant physical and social
factors implicated in a patient complaint if he or she is cut off
when trying to tell the full story of the reasons for seeking medical
care. The doctor may interrupt in order to zero in on a diagnosis,
but in so doing may miss a pertinent social or emotional factor
relevant to both diagnosis and appropriate treatment. Moreover, it
seems likely that the elderly, given the fact that they may take
longer to explain or give a history, are more subject to interrup-
tions and to the effects of an authority-based style of provider
practice (Haug & Ory, 1987).

Conventional belief would assign the more affective mode to
women in the tensions between technical and affective styles of
interaction with patients. Women are reported to be more sensitive
to relationships and understanding of others' problems and feelings
(Cartwright, 1972); thus, they are better able to establish rapport
with the elderly who frequently endure a mix of physical, emotional
and social problems. This stance is customarily applied more to
nurses than to physicians or dentists, but data to confirm or dis-
confirm the differential quality of the relationships are missing.
One study concerning task versus socio-emotional behaviors of physicians *vis-a-vis* their patients found that an affective style is a
persistent trait of certain male practitioners (Hall, Roter, & Katz,
1987), suggesting variations within a gender may be as significant
currently as that between genders.

The effect of gender on the shifting relationships between
patient, accompanying family member, and provider must be specu-
lative, since there are no data. Given that it is usually a wife or
daughter who goes along to a doctor or, more rarely, to a dentist,
it might be that she would be more likely to form a coalition if the
provider were also female. In the hospital, where many transactions
with nurses occur, a same gender dyad may also develop, particularly
if the elderly patient is a male. In any event, this can only be
guess work in the present state of our knowledge.

Whether women providers are more respectful toward the elderly and less disparaging than men is also uncertain. Ageist attitudes
have been found both in the predominantly female profession of
nursing and among the predominantly male physicians and dentists.
Dislike of geriatric nursing has been noted among female student
nurses (Kart, Metress, & Metress, 1978), and prejudicial attitudes
among male medical students about elderly women were actually
stronger than their racist views in one survey (Spence, Feigenbaum,
Fitzgerald, & Roth, 1968). Female medical students were slightly
less likely to stereotype the elderly negatively than were male
medical students (Belgrave, Lavin, Haug, & Breslau, 1982). It is
really impossible to confirm whether men or women are the worst
offenders when it comes to disparaging the elderly.

ETHNICITY

This author does not claim to be knowledgeable concerning ethnicity and provider-patient relationships, an area in which the comparative data are even more difficult to come by than those on gender. Only 2.2% of physicians are Black (Stoeckle, 1987), and supposedly a similar minority of dentists, and even fewer in either profession are Hispanic. On the other hand, the percentage of Black professional nurses, although still a minority, is somewhat larger (7%, personal communication). Most patients of Black doctors are themselves Black, but because of the paucity of practitioners of their own race many perforce must seek treatment from whites.

Ethnicity and professional-patient relationships. The quality of the physician-patient, or dentist-patient relationship, when ethnicity is incongruent is also complicated by class differences. Black elderly are likely to have had a limited education and to have been relegated to the most menial jobs with inadequate incomes. Aged Hispanics are apt to have suffered from the same deficits. Additionally, they may consider their command of English to be inadequate. In one current study of persons with arthritis, old Hispanics explicitly told Spanish speaking interviewers they avoid going to the doctor because of the language problem. Accordingly, provider-client relationships will be problematic not only because of ethnic disparity, but because of probable differences in social class, with attendant variations in language, response styles and cultural beliefs about illness (Haug & Ory, 1987).

One consequence may be hesitancy on the part of minority patients concerning self-disclosure in the absence of trust in the provider of a different race and class. Holding back may take the form of *"sullen reserve or even misleading friendliness, smiling and talkativeness"* (Santos, Hubbard, & McIntosh, 1983: 156). Elderly Black patients may be stereotyped by doctors as "happy," despite the fact that they are dissatisfied with their care, but believe they have no choice except to take what they can get (Satcher, 1973). Also compounding the effects of ethnicity, may be differences in religious beliefs, particularly with respect to supernatural or magical rather than medical reasons for illness and preferred method of treatment.

Whether Black or Hispanic providers and patients prefer a superordinate or an egalitarian mode of interaction is unknown. It is not even clear whether ethnic minorities would lean toward a technical or an affective style of giving or receiving care. One can certainly speculate, without much chance of disagreement, that preserving respect and dignity is particularly important for these groups of elderly, many of whom have had to deal with disparagement through much of their lives.

OTHER CHARACTERISTICS

Social class differences have been invoked in a number of studies of doctor-patient relationships, particularly with respect to patient compliance with professional regimen. Differences in socioeconomic status, most marked between physicians or dentists and their patients, have been blamed for failures in communication and understanding. In these analyses, class becomes a convenient surrogate for the effects of variations in education, income, and belief systems between providers and clients. The effect of religion on such belief systems, and consequently on interactions in medicine, nursing and dentistry, has never, to this author's knowledge, ever been studied. It is hardly likely that it is irrelevant. At a minimum, members of some religious groups might well be more comfortable with practitioners of their own faith who would understand the various behavioral restrictions to which they are committed. A Catholic doctor, after all, might inadvertently recommend the advantages of shellfish in the diet to an Orthodox Jew.

A word must be said about the location of any therapeutic transaction. There is a marked difference in the meaning of a relationship depending upon whether it occurs in a practitioner's office, in the patient's home, in a hospital, or a nursing home. If the interaction is to be in a doctor's office, the patient controls the decision whether or not to visit or keep an appointment. In a hospital or nursing home, the patient has virtually no control over physician visits, which can occur at any time--day or night. Sometimes, if the patient is an interesting case, or even if the condition is routine, the elderly patient will be seen by a coterie of medical students along with the doctor. The power of the patient therefore varies according to the organizational context of the relationship (Goss, 1981). After all, if dissatisfied, an ambulatory patient can switch doctors, which is an economic threat in a client dependent practice (Freidson, 1970), a power that is less easily exercised while in a hospital or nursing home.

Home visits by physicians have been exceedingly rare in the recent past. They may gradually be on the rise, in light of the current attempts to keep the seriously ill or frail and disabled elderly in their homes and out of institutions as long as possible as a cost containment measure. Although some may be brought to an office or outpatient clinic, others will need to be seen at home. Persons who are ill, perhaps particularly the elderly, are anxious and vulnerable. Being seen at home in their own familiar surroundings provides patients a sense of security not found on the doctor's turf, the office (Zola, 1986), and even fosters a feeling of control over the situation. Although the effect on the quality of the relationship can only be surmised, in the absence of empirical data, the empowerment of the patient is a significant factor in any treatment relationship.

It is worth noting that interactions with nurses occur almost entirely in institutions. Private duty and visiting nurses do work in the home on occasion, and there are nurses in private office practice, particularly nurse practitioners and midwives, but these are still very rare. Accordingly, patients hardly ever have the power to seek or decline nursing services. In the hospital and nursing home the nurse, backed by bureaucratic rules and regulations, controls the time and place, and usually the content, of any interaction. In some ways patients are most powerless in dealings with nurses because, as patients, they are ill and confined to an institution.

CONCLUSIONS AND FURTHER SPECULATIONS

By this time, the reader must be glad that this paper did not try to cover all 18 million variations in provider-patient relationships when the patient is elderly. But perhaps some inkling of the potential effects of age and gender and the character of the transaction has become clearer, even though it has been possible to give only passing reference to the effects of ethnicity, class, and religion. It is sadly apparent that there is a large gap in knowledge concerning what occurs when elderly patients meet with doctors, dentists or nurses, not to mention the nature of interactions between the elderly and physical therapists, pharmacists, social workers or other types of providers.

Moreover, the effects of interaction modalities on quality of care and health outcomes of the elderly are largely unknown as are their consequences for utilization or failure to utilize professional services. Indeed this dearth of research has recently been recognized by the National Institute on Aging, which in early 1987 issued an RFP for the study of formal health care, including a subsection on provider-patient relationships.

Forecasting the future direction of therapeutic relationships is problematic at best. It is possible to predict some changes in the characteristics of future elderly since they are living today. By the year 2000 there will be nearly 32 million persons over 65, almost half of whom will be over 75 (Verbrugge, 1983). The young-old in this population, born in the early thirties, are the young people of the fifties and sixties who generally enjoyed more education and showed more skepticism of authority in general and of physician authority in particular (Haug & Lavin, 1981), than was manifested among older cohorts at the time.

This on-coming cohort of elderly persons has more knowledge of health issues gained from their experiences with modern medicine and the media. There is some evidence that they are more willing to identify bodily changes as illness than older cohorts (Shorter, 1985) while, at the same time, they are more alienated and less trustful of physicians than was the case in the past. Less is known about physicians of the future, although some younger ones currently have demonstrated greater willingness than older practitioners to accept a more egalitarian relationship with their patients.

Other relevant changes are also taking place. New forms of medical practice, like urgent care centers, HMO's and PPO's, are becoming common. They are now advertising for elderly patients. How will provider competition or method of payment affect the provider-patient relationships? The drug revolution of the forties--the discovery of penicillin and the antibiotics--substantially changed interactions as patients came to expect magic bullets not previously available for their ailments, and physicians spent less time exploring the psychosocial issues and personal stresses underlying physical complaints as they wrote prescriptions and ordered tests, perhaps to excess (Shorter, 1985). Furthermore, will new technological discoveries, like the ability to transplant organs, or to computerize diagnosis and treatment, transform old relationship styles?

Finally, will the sheer relative number of the elderly, up to 20% and more of the population in the United States by the end of the century, change attitudes so that ageism fades, or will it become exacerbated as a result of intergenerational pressures? The answer may well be relevant to one common need of the elderly today. As stated in an earlier study on this same issue:

> In a society where ageism is still widely prevalent, older people's sense of self-worth is undermined by ageist attitudes or actions in many everyday situations. Thus, when older people are especially anxious and vulnerable, as they often are in health care interactions, it becomes more important than ever to have the old patient's dignity and self-worth reinforced by a caring, empathetic provider, rather than diminished by one whose approach is abrupt, impatient, and purely technical (Haug & Ory, 1987).

With more old people and greater competition for their business, practitioners, for economic reasons, are apt to treat their elderly clients more deferentially, with consideration and respect. Should that occur, the demographic imperative will have had a positive effect. ▧

REFERENCES

Belgrave, L.L., Lavin, B., Breslau, N., & Haug, M.R. Stereotyping of the aged by medical students. *Gerontology and Geriatric Education*, 1982, 3(Fall), 37-44.

Bloom, S.W., & Speedling, E.J.   Strategies of power and dependence in doctor-patient exchanges. In M.R. Haug (Ed.), *Elderly patients and their doctors.* New York: Springer, 1981.

Bluestone, N.R.   The future impact of women physicians on American medicine. *American Journal of Public Health, 1978,* 68(8), 760-763.

Boston Women's Health Book Collective.   *Our bodies, ourselves.* New York: Simon & Schuster, 1973.

Butler, R.N.   *Why survive? Being old in America.*   New York: Harper & Row, 1975.

Cartwright, L.K.   Personality differences in male and female medical students. *Psychiatry Medicine,* 1972, 3, 213-218.

Church, O.M., & Poirier, S.   From patient to consumer: From apprentice to professional practitioner. *Nursing Clinics of North America,* 1986, 21(1), 99-109.

Ciliberto, D.J., Levin, J., & Arluke, A.   Nurses' diagnostic stereotyping of the elderly: The case of organic brain syndrome. *Research on Aging,* 1981, 3(3), 299-310.

Claus, L.M.   Dental student attitudes towards the elderly and training in geriatric dentistry. *International Dental Journal,* 1982, 32(4), 371-378.

Coe, R.M., & Prendergast, C.G.   The formation of coalitions: Interaction strategies in triads. *Sociology of Health and Illness,* 1985, 7, 236-247.

Collett, H.A.   Influence of the dentist-patient relationship on attitudes and adjustment to dental treatment. *Journal of the American Dental Association,* 1969, 79, 879-84.

Conant, E.B.   Addressing patients by their first names. [Letter to the editor]. *New England Journal of Medicine,* Jan., 1983, 226.

Dall, C.E., & Gresham, L.   Promoting effective drug-taking behavior in the elderly. *Nursing Clinics of North America,* 1982, 17(2), 283-291.

Eisenberg, J.M.   Physician utilization: The state of research about physician's practice patterns. *Medical Care,* 1985, 23(5), 461-483.

Ettinger, R., Beck, J.D., & Jakobsen, J.   The development of teaching programs in geriatric dentistry in the United States from 1974 to 1979. *Special Care in Dentistry,* 1981, 1, 211.

Evangela, M.   The influence of family relationships on the geriatric patient: The nurse's role. *Nursing Clinics of North America,* 1968, 3(4), 653-662.

Ford, C.V., & Sbordone, R.J.   Attitudes of psychiatrists toward elderly patients. *American Journal of Psychiatry,* 1980, 137, 5, 571-575.

Freidson, E.   *Professional dominance.*   New York: Antherton Press, 1970.

Goss, M.E.W.   Situational effects in medical care of the elderly: Office, hospital & nursing home.   In M.R. Haug (Ed.), *Elderly patients & their doctors.* New York: Springer, 1981.

Green, C.P.   How to recognize hostility & what to do about it. *American Journal of Nursing,* 1986, 86(1), 1230-1234.

Hall, J.A., Roter, D.L., & Katz, N.R.   Task versus socioeconomic behaviors in physicians. *Medical Care,* 1987, 25(5), 399-412.

Haug, M.R.   Doctor patient relationships and the older patient. *Journal of Gerontology,* 1979, 34(6), 852-860.

Haug, M.R., & Folmar, S.   Longevity, gender and life quality. *Journal of Health and Social Behavior,* 1986, 27(4), 332-345.

Haug, M.R., & Lavin, B.   Practitioner or patient - who's in charge? *Journal of Health and Social Behavior,* 1981, 22, 212-229.

Haug, M.R., & Ory, M.   Issues in elderly patient-provider interacttion. *Research on Aging,* 1987, 9(1), 3-44.

Haug, M.R., & Sussman, M.B.   Professional autonomy and the revolt of the client. *Social Problems,* 1968, 17, 153-161.

Heins, M.   Career and life patterns of women and men physicians. In E.C. Shapiro & L.M. Lowenstein (Eds.), *Becoming a physician.* Cambridge, MA: Ballinger, 1979.

Hooper, E.M., Comstock, L.M., Goodwin, J.M., & Goodwin, J.S.   Patient characteristics that influence physician behavior. *Medical Care,* 1982, 20(6), 630-638.

Ingham, R., & Fielding, P.   A review of the nursing literature on attitudes toward old people. *International Journal of Nursing Studies,* 1985, 22(3), 171-181.

Kandelman, D., & Lepage, Y.   Demographic, social and cultural factors influencing the elderly to seek dental treatment. *International Dental Journal,* 1982, 32(4), 360-370.

Kart, C.   Experiencing symptoms: Attributions and misattributions of illness among the aged. In M.R. Haug (Ed.), *Elderly patients and their doctors.* New York: Springer, 1981.

Kart, C.S., Metress, E.S., & Metress, J.F.   *Aging and health: Biologic and social perspectives.* Reading, MA: Addison-Wesley, 1978.

Katon, W., & Kleinman, A.   Doctor-patient negotiation and other social science strategies in patient care. In L. Eisenberg & A. Kleinman (Eds.), *The relevance of social science for medicine.* Boston: D. Reidel, 1981.

Keeler, E.B., Solomon, D.H., Beck, J.C., Mendenhall, R.C., & Kane, R.L.   Effect of patient age on duration of medical encounters with physicians. *Medical Care,* 1982, 20(11), 1101-1108.

Kiyak, H.A., Milgrom, P., Ratener, P., & Conrad, D.   Dentist's attitudes toward and knowledge of the elderly. *Journal of Dental Education,* 1982, 46(5), 266-273.

Kleinman, M.B., & Clemente, F.   Support for the medical profession among the aged. *International Journal of Health Services,* 1976, 6(2), 295-299.

Kosberg, J.I.   The importance of attitudes on the interaction between health care providers & geriatric populations. *Interdisciplinary Topics in Gerontology,* 1983, 17, 132-143.

Lavin, B.   Gender and the medical student.   Paper presented at the North Central Sociological Association Meeting, Detroit, MI, May, 1982.

Lavin, B., Haug, M.R., Belgrave, L.L., & Breslau, N.   Change in student physicians' views on authority relations with patients. *Journal of Health and Social Behavior*, 1987, 28, 258-272.

Levinson, R.M., McCollum, K.T., & Kutner, N.G.   Gender homophily in preferences for physicians.   *Sex Roles*, 1984, 10(5/6), 315-325.

MacEntee, M.I.   The dentist and the older patient: A review of the relationship.   *Journal of Canadian Dental Association*, 1984, 9, 675-678.

Marshall, V.W.   Physician characteristics and relationships with older patients.   In M.R. Haug (Ed.), *Elderly patients and their doctors.* New York: Springer, 1981.

Moses, D.V.   The older patient in the general hospital.   *Nursing Clinics of North America*, 1967, 2(4), 705-714.

Pendelton, D.A., Brouwer, H., & Jaspars, J.   Communication difficulties: The doctor's perspective. *Journal of Language and Social Psychology*, 1983, 2, 17.

Riley, M.W.   The changing older woman: A cohort perspective.   In M.R. Haug, A.B. Ford, & M. Sheafor (Eds.), *The physical and mental health of aged women.* New York: Springer, 1985.

Robert Wood Johnson Foundation.   *Medical Practice in the United States - A special report.* Princeton: Author, 1981.

Rosow, I.   Coalitions in geriatric medicine. In M.R. Haug (Ed.), *Elderly patients and their doctors.* New York: Springer, 1981.

Santos, J.F., Hubbard, R.W., & McIntosh, J.L.   Mental health and the minority elderly.   In L.D. Breslau & M.R. Haug (Eds.), *Depression and aging: Causes, care & consequences.* New York: Springer, 1983.

Satcher, D.   Does race interfere with the doctor-patient relationship?   *Journal of American Medical Association*, 1973, 223(13), 1498-1499.

Schoen, E.A.   Clinical problem: The demanding, complaining patient. *Nursing Clinics of North America*, 1967, 2(4), 715-724.

Shapiro, E.C., & Jones, A.B.   Women physicians and the exercise of power and authority in health care. In E.C. Shapiro & L.M. Lowenstein (Eds.), *Becoming a physician: Development of values and attitudes in medicine.*   Cambridge, MA: Ballinger Publishing Company, 1979.

Shorter, E.   *Beside manners: The troubled history of doctors and patients.* New York: Simon & Schuster, 1985.

Spence, D.L., Feigenbaum, M., Fitzgerald, F., & Roth, J.   Medical student attitudes toward the geriatric patient. *Journal of the American Geriatrics Society*, 1968, 16(9), 976-983.

Stewart, M.   Patient characteristics which are related to the doctor-patient interaction. *Family Practice*, 1983, 1(1), 30-36.

Stoeckle, J.D.   Introduction. In J.D. Stoeckle (Ed.), *Encounters between patients & doctors*. Cambridge, MA: MIT Press, 1987.

U.S. Senate Special Committee on Aging. *America in transition: An aging society*, 1984-85 edition, Serial #99-B.   Washington, D.C.: Government Printing Office, 1985.

Verbrugge, L.M.   Women and men: Mortality and health of older people. In M.W. Riley, B.B. Hess, & K. Bond (Eds.), *Aging in society: Selected reviews of recent research*.   Hillsdale, NJ: Lawrence Erlbaum Associates, Publishers, 1983.

Waitzkin, H., & Stoeckle, J.D.   Information control and the micro-politics of health care: Summary of an ongoing research project. *Social Science & Medicine*, 1976, 10, 263-276.

Weisman, C.S., Levine, D.M., Steinwachs, D.M., & Chase, G.A.   Male and female physician career patterns: Specialty choices & graduate training. *Journal of Medical Education, 1980*, 55(Oct.), 813-825.

Weisman, C.S., & Teitelbaum, M.A.   Physician gender and the physician-patient relationship: Recent evidence and relevant questions. *Social Science & Medicine*, 1985, 20(11), 1119-1127.

West, C.   *Routine complications: Troubles with talk between doctors and patients*. Bloomington: Indiana University Press, 1984.

Winslow, E.H.   The role of the nurse in patient education. *Nursing Clinics of North America*, 1976, 11(2), 213-222.

Yuen, F.K.H.   The nurse-client relationship: A mutual learning experience. *Journal of Advanced Nursing*, 1986,11(5), 529-533.

Zola, I.K.   Reasons for noncompliance and failure of the elderly to seek care.   In R. Moskowitz & M.R. Haug (Eds.),   *Arthritis and the elderly*. New York: Springer, 1986.

⧚ 18 ⧚

THE WALKING PATIENT AND THE REVOLT OF THE CLIENT: IMPETUS TO DEVELOP NEW MODELS OF PHYSICIAN-PATIENT ROLES

Betty E. Cogswell

INTRODUCTION

For over two decades investigators have been aware that earlier models of the physician-patient role required modifications, revisions, or replacement by alternative models.[1] In these early models, physicians are viewed as holding authority and power in an asymmetrical relationship with patients and their families. According to Parsons (1951), physicians' power is based on having what patients want and need. Patients are passive and dependent recipients of care. Physicians are awarded professional prestige and maintain situational authority, thus their advice should be followed because "*doctor knows best.*"

One deficiency of the Parsons model of the sick role is that he speaks only to expectations for behavior and does not deal with the actual behavior (Bloom & Wilson, 1972; Freidson, 1961). Thus, the model does not benefit from insights gained in physician-patient encounters that provide data for correcting theory. Another deficiency is that Parsons' model, as well as those proposed by other theorists (Szasz & Hollender, 1965), is conceived from the physicians' perspectives, neglecting the perspectives of the patients.

Evidence suggests that these physician-patient models are still the models dominating theory, practice and research. Patients as well as physicians continue to support a physician-controlling model. For example, numerous studies of patients in the past 20-30 years tend to reflect issues important to physicians such as research on patient compliance, reflecting what physicians, rather than patients, think is important. This stance led Haug and Sussman (1969:154) to

243

observe that: *"This idealized model is not now, if it ever was, iso-morphic to reality."*

Sussman, Caplan, Haug, and Stern (1967), in a study of the walking patient, and Haug and Sussman (1969), in an article on the revolt of the client, demonstrated the need for changes in the traditional portrayal of the physician-patient role to bring this conceptual framework into closer relationship with reality. The traditional model, first posed during an era dominated by in-patient health care, did not accurately represent physician-patient relationships in outpatient settings.   In addition, the early physician-patient role models were developed during a time when patients with acute illnesses, rather than those with chronic conditions, constituted the majority of physicians' practices (Sussman *et al.*, 1967).

The revolt of the client is a growing reality in the 1980s, but even more important, this phenomenon is a reminder that those in the field of medicine as well as investigators who study this field have been remiss too long in not studying the physician-patient role from the "patients as consumers" perspectives.

In this chapter we will explore: the need to develop new physician-patient models to reflect more closely the changing and the real world; the need to encourage research to obtain the necessary data for theoretical development of physician-patient role models from the consumers' perspectives to redress the bias of previous models built on physicians' views; the need to develop more humanized, effective, higher quality care; and the need for physicians rather than patients to act as change agents in implementing new physician-patient roles.

BRINGING MODELS INTO CLOSER RELATIONSHIP
WITH THE REAL WORLD

Delivery of health care is hampered by models of physician-patient roles that inadequately connect with the real world. Blumer (1953:54) notes:

Theory is of value in empirical science only to the extent to which it connects fruitfully with the empirical world. Concepts are the means, and the only means of establishing such connection, for it is the concept that points to the empirical instances about which a theoretical proposal is made. If the concept is clear as to what it refers, then sure identification of the empirical instances may be made. With their identification, they can be studied carefully, used to test theoretical proposals and exploited for suggestions as to new proposals. [Only] with clear concepts [can] theoretical statements . . . be brought into close and self-correcting relations with the empirical world.

Szasz and Hollender (1956) reach a somewhat closer co
with the empirical world in their three-model construct of ph
patient relationships. The first level of this model, a
passivity, is clinically appropriate to the unconscious or acutely
traumatized patient. The second level, guidance-cooperation, is
appropriate for acute disorders. The third level of the model,
mutual participation, is appropriate for management of chronic illness and for psychoanalysis. This is considered to be an improvement over Parsons's model because it recognizes that patients, especially those with chronic diseases, cannot remain passive but
must actively participate in their own health care. The Szasz-
Hollender model, however, according to Bloom and Wilson (1972:324):

> . . . parallels Parsons in its asymmetry, leaning heavily on
> the parent-child analogy, with similar implications for the
> necessity of status differences between the roles, with the
> professional in firm control.

A model of mutual participation between doctors and patients
would understandably emerge from experiences with chronically ill
patients. The very nature of treatment of chronic disease guides
physicians away from a controlling role and toward a participatory
role with patients, because patients with chronic diseases (except for
acute episodes) are treated in ambulatory settings. Out-patient
physicians have neither the organizational supports nor constraints
inherent in in-patient (hospitalized) services that permit them to
order care. Out-patients are responsible for their own daily care,
so physicians teach patients and their families, rather than act on
them. Physicians working predominantly with chronically ill and
disabled patients, and more recently those providing preventive and
health promotion care, have come to recognize the necessity of the
active involvement of the patient. Even though some physicians
speak out for less controlling and more democratic postures toward
patients (Coles, 1986; Siegel, 1986), doctors who assume controlling
stances are still very much in evidence.

In very broad terms, historical events have changed the way
physicians and patients relate to one another. From the early
history of the United States until well into the twentieth century, a
pre-specialization form of physician-patient role prevailed. During
this period hospitals were few and distant, and medical care and
recovery frequently occurred in patients' homes. Physicians had to
rely on patients and their families to carry out medical instructions,
since there were few trained ancillary personnel. Physicians' directions had to be clear and explicit for lay caretakers to be effective,
and instructions had to include contingencies--if "A" occurs, do this;
if "B" occurs, do that.

Beginning in the 1930s and 40s a new specialized physician-
patient relationship began to develop in tandem with rapid progress

in scientific medicine and increased availability and use of hospitals. In hospitals patients are under 24-hours-a-day care by trained ancillary personnel. Physicians can give one specific order and change it if, or when, contingencies arise. There is no need to simplify orders so that patients and their family members can understand, because nurses are familiar with physicians' language. In addition, hospitals provide physicians with the organizational supports and constraints to maintain control of patients, which reduces physicians' needs to communicate directly with them.

In the 1940s, with a rise in outpatient visits and with increases in the incidence and prevalence of chronic disease, physician-patient relationships began to develop in a way that required patients to be active participants in their own care. In contrast to the era of pre-specialization, physicians now possessed a large body of scientific knowledge and had mastered new and more sophisticated techniques. This knowledge and techniques were learned in inpatient settings that were not conducive to physicians' developing the interpersonal skills necessary to care for out-patients. Caring for walking patients necessitates their participation. Ambulatory patients are beyond daily surveillance and help by physicians and trained staff. Therefore, they and their family members perform much of the care. To be effective, patients and their families need sufficient skills and information in order to monitor the patients' condition, recognize relevant changes, and to be able to carry out the appropriate actions for an array of likely contingencies. For walking patients, it is difficult if not impossible for high quality care to occur without their understanding, cooperation, and active involvement.

PHYSICIAN-PATIENT ROLES FROM THE CONSUMERS' PERSPECTIVES

During these years of rapid scientific advances in medicine and the rise of specialization, numerous societal changes distanced patients from physicians and reduced patients' influence in physician-patient encounters. Since 1910, reactions to the Flexner report have led to the reorganization of medical education and the proliferation of medical specialties (Cogswell, 1981; Cogswell, Aluise, Shahady, & Thomas, 1986). Concomitantly, health care institutions and professions became increasingly independent of the public and care became organized according to professional rather than lay standards (Freidson, 1970; Starr, 1982). This period gave rise to traditional models of the physician-patient role.

Furthermore, in recent decades professional disregard of consumers' standards for care has been encouraged and sustained by a shortage of primary care physicians, mediation by third parties, bureaucratic service delivery, continually advancing technology, and, most recently, efforts toward cost containment and alternative prepaid plans (Bloom & Summey, 1976; Stephens, 1982). Health care professionals and policymakers who recognize and seek to correct this imbalance in provider-consumer relationships are hampered by a lack of systematic knowledge about the way the public sees physician-patient roles.

To improve the practice of medicine, there is a need to develop theoretical models of physician-patient relationships that fruitfully connect with the empirical world. To theoretically ground these models (Glaser & Strauss, 1967) requires, at a minimum, studies of patients'-consumers' perspectives on this role as well as studies of physicians' perspectives. Indeed, if we require a full-blown view of these phenomena, we should take Freidson's advice (1961) and study this role from the points of view of patients, their family members, physicians, nurses, and other ancillary personnel. One would expect the perspectives of each of these types of actors to be different. Models based on data about patients' perspectives should help physicians to understand patients better and to modify their relationships to meet concerns and priorities thereby rendering client revolts less volatile. Haug and Sussman (1969; 153) very astutely point out that various publics are questioning professionals' claims:

> . . . to special knowledge and the humanitarian ethos and challenging institutional delivery systems, either because they are inadequate or because they exceed appropriate bounds.

When they speak of revolt, however, they are talking about patients-clients, not as individuals, but as a group of people who organize to protest. Patients in general are less able than physicians to take organized group action. Among patients, group formation is most apt to develop in situations where patients know and have regular contact with each other, as in neighborhood clinics or total institutions such as mental hospitals (Goffman, 1961a), residential rehabilitation centers, or experimental wards.

However, most patients have minimum access to each others' experiences as patients. Primary health care occurs within the privacy of physician-patient encounters. Thus, people are socially isolated from each others' experiences as patients. Because of this social isolation, patients are an aggregate of individuals rather than an organized group and they are relatively powerless in negotiating new role definitions with physicians. Loosely organized groups, such as the women's health movement, have brought about some improvements in response to their protests, but have been only partially successful in achieving their priorities (Cogswell & Arndt, 1980). Therefore, redress of imbalances in physician-patient roles may rest on the actions of leaders within the field of medicine who champion patients' causes and on social science investigators who directly address patients' perspectives.

In one of the first systematic studies of patients' perspectives on medical practice, Freidson (1961: 209) clearly identifies the quality of the physician-patient relationship as a salient aspect of care.

> *Most patients naturally desired what they thought to be good technical care, but they insisted nonetheless that without personal interest the practitioner could not use his full competence.*

Freidson concludes by calling for the empirical study of forms of practice to discover behavior of physicians that interferes with successful patient care. Fifteen years later, Pratt (1976) and Kelman (1976) note that "professional" assumptions dominate a substantial body of research on the physician-patient relationship. Pratt emphasizes the need for a comparable literature based on consumers' views in order to fully assess the relative effectiveness of different models of health care delivery. More recent research comparing physicians and patients (Cogswell *et al.*, 1986, Gillette, Kues, Harrigan, & Franklin, 1986; Haug & Lavin, 1983; Hyatt, 1980; Scammon & Kennard, 1983; Warner, 1981) suggests that doctors maintain a perspective substantially different from that of their patients, and are generally poor judges of patients' perspectives. Thus, one may assume that physicians do not have even implicit models of physician-patient roles that reflect patients' perspectives sufficiently to serve as frameworks for organizing insights about the patients they encounter on a day-to-day basis.

MORE HUMANISTIC AND HIGHER QUALITY CARE

It is suggested that physician-patient roles based on equality, cooperation, and mutual participation lead to more humanistic and higher quality health care than the earlier and still more predominant superordinate-subordinate roles. Mounting evidence from disparate sources ranging from industry to service programs, and from developing to developed nations, indicates that mutual active participation in decision making and control of common endeavors yields results of higher productivity and quality than unilateral control by experts (Whyte, 1984).

Service delivery styles and management of workers, as Whyte notes, are less effective when based on paradigms built on these four assumptions:

* That professional experts know the "one best way;"

* That improvement should be unilaterally initiated by experts;

※ That experts' recommendations, if applied, yield favorable results; and

※ That tight control of an endeavor should remain in the hands of experts.

Autocratic styles of practice and management are less effective, reduce quality and productivity, and dehumanize subordinates. Delivery of humanized care is difficult if not impossible in programs designed and administered without consideration of consumer-patient standards.

A dimension of the professional attitude set forth by Parsons (1951) is that physicians should be oriented toward universalism. This concept refers to practitioners treating all patients the same way by viewing them as equal (Bloom & Wilson, 1972). Robert Coles, a physician, takes a contrary position and perhaps better reflects patients' priorities:

The doctor treats one person, then another, aware all the time (one hopes) that variation is a great constant of the work he or she does--those aspects of our individuality, those idiosyncracies, those personal habits and beliefs and wishes and worries that distinguish each of us from the other (1986:2125).

Seeing many patients, each different from the other, requires variations in approach. Mutual participation between doctor and patient allows each patient more latitude to speak for him/herself and be heard by the physician. For physicians to improve the quality of health care delivery, we need research on consumers' experiences from their own perspectives which reflect different backgrounds and life situations. Models of physician-patient relationships built upon consumers' own perspectives would let physicians take advantage of knowledge about likely behavior patterns without being surprised by deviations from modal patterns. In fact, unexpected patient behavior should be a signal to physicians that there is a need to learn more about that patient. Such models, especially those based on qualitative research, can help identify the broad features of consumers' perspectives that tend to be the same for many patients, but also can alert practitioners to the types of specifics that may be different for each individual.

Physicians' and patients' understanding of each other is based on their appraisals of reciprocal, not unidirectional interactions. Only by empathetically placing themselves into their patients' roles can physicians understand patients' actions and expectations. Qualitative studies of patients, because emphasis is on people not traits or attitudes, could sensitize physicians to patients' perspectives and could enhance physicians' ability to *"walk in patients' shoes."* As an example of this process, Siegel describes positive and self-fulfilling changes in his relationships with patients as a result of opening up and revealing himself to them as a person. He attributes his success in treating cancer patients to his acting like a person instead of acting as he was trained which was *"to think my whole job was doing things to people in a mechanical way to make them better, to save their lives"* (1986:11).

Physicians' understanding of patients is a necessary precondition for humanized care. Understanding enhances the quality of the physician-patient relationship, leads to mutual respect between doctors and patients, enlists patients' cooperation and active involvement in carrying out physicians' advice, and increases the possibility that doctors and patients will work together creatively in a spirit of cooperation to solve patients' health care problems. Human understanding of patients has the potential for increasing the quality of both processes and outcomes of care.

Understanding patients, however, cannot be gained merely by studying aspects of patients (attitudes, values) or patients as units rather than people, nor can it be gained outside the context of physician-patient roles. Rather, questions for investigation should focus on patients' views of physician-patient interactions.

PHYSICIANS AS CHANGE AGENTS

Siegel (1986), through personal experience with his own paients, realized that healing occurred not from his doing something to patients but by his participation in their recovery. He reports that:

> *The most important kind of assertiveness a patient can demonstrate is the formation of a participatory relationship with the doctor* (p. 172). . . *Participation in the decision-making process, more than any other factor, determines the quality of the doctor-patient relationship. The exceptional patient wants to share responsibility for life and treatment, and doctors who encourage that attitude can help all their patients heal faster* (p. 51).

To replace asymmetrical physician-controlled/patient relationships with participatory relationships requires changes in the behavior of physicians and patients, as well as changes in the organizational structure of health care settings. If these changes come about, and social pressures are certainly in this direction, who will be the agents of change? Patients as individuals can be assertive in their encounters with physicians or can seek a new physician when their present one does not meet their expectations. Furthermore, they can organize to protest disagreeable styles of practice or to bargain for changes they would like to occur. However, patients organizing patients' rights associations is unlikely to occur on a widespread basis.

For a number of reasons, physicians are in a better position than patients to initiate changes in their role relationships; but physicians also have constraints. The most inhibiting constraint is the absence of alternative models of physician-patient roles. A physician-defined model of these roles persists primarily as a result of four factors.

First, physicians have repeated, daily encounters with multiple patients, thus their experience with patients is considerably greater than patients' experiences with doctors. In contrast, patients see physicians infrequently. Since physicians may be acquainted with several hundred patients, while patients tend to know only a few physicians, physicians have a much larger base of experience from which to gain insights about the physician-patient relationship. Their cumulative experience, however, may be far from corrective in treating patients as people, because of organizational mandates and constraints that give them little time for reflection on and analysis of the data they possess. Physicians' insight and sympathy tend to become more alienated and rationalized because of pressures of time and economics and boundaries of specialized knowledge and roles (Sarason, 1985). Furthermore, physicians who use traditional controlling models to view, organize, and assess their relationships with patients have little rationale or motivation to change.

Second, not only do physicians see patients on their turf (clinics, offices, hospitals), since home visits are rarely made, they are also the hosts and patients their "guests." Only very assertive patients would try to invoke their own rules of behavior rather than accept those of the physician/hosts. Therefore, physicians are better able to initiate and invite changes in the relationship.

Third, physicians are organized into professional groups that facilitate communication as well as group efforts to effect change. Thus, they have greater opportunities to act as a group or subgroup of kindred spirits within the field of medicine. Although individual physicians have written and acted to champion the cause of patients, there is increasing evidence that organized groups also are becoming concerned about patients' views.[2]

Finally, physicians have common socialization experiences passing as cohorts through medical school and residency. Their socialization into the role of physician is formal and structured with designated teachers and organized schedules occurring within formal social organizations (medical schools, hospitals). Patients, however, must learn their role informally without the benefit of designated teachers or structured curriculum or time. Thus, patients' definitions of their role are apt to be varied, amorphous, tentative, and often ambiguous. In contrast, the lengthy formal socialization of physicians leads to their having much clearer definitions of their role.

Although there are always variations in role definition among role occupants, physicians who undergo formal socialization are more apt to have less variability in role definitions than patients whose socialization into the role of patient is informal and unstructured. As a result, physicians play their roles with much more certainty and assurance than patients play their roles. Furthermore, for people who are physicians, the role is major; but for most people who are patients, the role is peripheral, played intermittently for brief periods throughout a lifetime. Physicians who draw a major portion of their identity from their role are apt to have more at stake in the way the role is played and more often are apt to exercise their role definitions to the detriment of patients' definitions.

Physicians' roles have changed, but socialization lags. An unfortunate by-product of the age of specialization is that the majority of socialization experiences for student physicians (medical students and residents) are with in-patients in hospital settings characterized by controlled environment. This physician-in-patient role is counterproductive for out-patient care.

Unfortunately, with the exception of residencies in family medicine, which include a sizeable portion of out-patient care, physicians in training receive minimal first-hand experience with out-patients/walking patients. Thus, new physicians entering practice are operating with training that is inappropriate for care of out-patients. During training, primary care physicians need sufficient in-patient experience to be able to manage their own hospitalized patients and to coordinate care with specialists, but they need considerably more out-patient experience to cope successfully with the majority of patients they will encounter in practice. Even physicians who subscribe to the value that patients are people rather than numbers or cases and behave accordingly, may find themselves lapsing into the specialization model of physician-patient roles simply because this has constituted the bulk of their training.

CONCLUSIONS

To develop models of physician-patient roles that could be useful as guides for medical practice first requires data and then conceptual analysis (Blumer, 1953). As Freidson (1961:190) noted:

Concepts have little value if they do not pick out salient features of reality, and in the case of the doctor-patient relationship the reality is sufficiently fluid to make it difficult to know exactly what concepts would be most useful.

In general, the fluidity of the physician-patient relationship can be categorized into three historical periods: the pre-specialization era of medicine where interpersonal relationships between physicians and patients were necessary for care; the specialization era, accompanied by the medical professions distancing itself from patients; and the current era where out-patient care of chronically ill people, occurring in tandem with continual new developments in scientific knowledge and technology, again requires physicians to depend heavily on interpersonal relationships with patients to achieve quality care.

Both social structure and the physical realities confronting physicians and patients set boundaries by limiting the alternatives for social interaction. Physical reality, which includes such factors as the absence or presence of hospitals or chronic disease, does not direct how roles are played, but merely places constraints on the number of alternatives for behavior (Cogswell, 1967), while the social structure, the framework within which interaction occurs, is influential in shaping situations in which people act. Participants make successive choices among the degrees and ranges of alternatives within the outer limits imposed by the social structure (Cogswell, 1968).

Therefore, the most advantageous way of developing new theoretical models of physician-patient relationships is to focus on social interaction, but to be constantly aware, and note the influences, of social structure and the nature of socially impinging physical issues. To be useful for physicians in their day-to-day practice of medicine, models should serve as a framework for ordering daily observations about the patients they see and to encourage and facilitate new insights into their relationships with patients. New models should connect more closely with the empirical world, and emphasize the patients' perspectives on their interactions with physicians.

There is evidence within the field of medicine that physicians are developing more participatory roles with patients. Where they exist, these role relationships should be studied to learn their nature, how they emerge, and whether or not they increase the quality of care. Fields such as business and industry, with an ethos of competition and profit, have found that mutually active participation in decision making and control of work by labor and management increases both the productivity and quality of their products (Whyte, 1984). Thus, it seems plausible that medicine, with an ethos of humanitarianism, would be a likely candidate to explore similar approaches to physician-patient roles. Subgroups within medicine, particularly family medicine, have made considerable headway in this direction in rhetoric if not completely in practice (Cogswell, 1981). Models of participatory physician-patient roles would provide physicians with a mechanism for organizing their own experiences, as well as rationales for taking leadership roles that enabled them to extend this approach to other segments of the health care field.

Given the increasing life expectancy, exploding frontiers in medical technology, and cost-containment concerns, patients and their families will be required to assume a participatory role in their health care across the life span. Therefore, it is incumbent upon the medical profession to assume the leadership as change agents. ▩

I wish to express my appreciation to Alfred O. Reid, Jr., of the Department of Family Medicine at the University of North Carolina at Chapel Hill, who through many discussions helped me clarify my thinking on these issues, to Lynn Igoe and Catheryn Brandon at the University's Carolina Population Center, for editorial suggestions and for word-processing assistance, respectively.

ENDNOTES

1. See Henderson (1935), Parsons (1951) and Szasz & Hollender (1956) for early models of the physician-patient roles, and Bloom (1963), Bloom & Wilson (1972), Cogswell *et al.*, (1986), Freidson (1961), Kelman (1976) and Pratt (1976) for a call for modifications, revisions or replacement of these earlier models.

2. Since 1982, the American Medical Association (AMA) has been conducting annual national surveys of physician and public opinions. The most recent survey, 1986, consistent with earlier surveys, found that the publics' image of physicians is largely negative (Harvey & Shubat, 1986).

REFERENCES

Bloom, S.W. *The doctor and his patient.* New York: Russell Sage Foundation, 1963.

Bloom, S.W., & Wilson, R.N. Patient-practitioner relationships. In H.E. Freeman, S. Levine, & L.G. Reeder (Eds.), *Handbook of medical sociology.* Englewood Cliffs, NJ: Prentice-Hall, 1972.

Bloom, S.W., & Summey, P. Models of the doctor-patient relationship: A history of the social system concept. In E. Gallagher (Ed.), *The doctor-patient relationship in the changing health scene.* New York: Geographic Health Studies, John E. Fogarty International Center for Advanced Study in the Health Sciences, 1976.

Blumer, H. What is wrong with social theory? In W.J. Filstead (Ed.), *Qualitative methodology: Firsthand involvement with the social world.* Chicago: Markham, 1970.

Blumer, H. *Symbolic interactionism: Perspective and method.* Berkeley: University of California Press, 1969.

Cogswell, B.E. Rehabilitation of the paraplegic: Processes of socialization. *Sociological Inquiry*, 1967, 37, 11-26.

Cogswell, B.E. Some structural properties influencing socialization. *Administrative Science Quarterly*, 1968, 13, 417-440.

Cogswell, B.E. Family physician: A new role in process of development. *Marriage and Family Review*, 1981 4(1-2), 1-30.

Cogswell, B.E., & Arndt, J.E. Women's health care: An overview. *Marriage and Family Review*, 1980, 3(3/4), 1-29.

Cogswell, B.E., Aluise, J.J., Shahady, E.J., & Thomas, K.K. Family medicine from consumers' perspectives. In W.J. Doherty, E.E. Christianson, & M.B. Sussman (Eds.), *Family medicine: The maturing of a discipline.* New York: Haworth Press, 1986.

Coles, R. Literature and medicine. *Journal of the American Medical Association*, 1986, 256, 2125-2126.

Flexner, A. *Medical education in the United States and Canada.* A report to the Carnegie Foundation for the Advancement of Teaching. Carnegie Foundation Bulletin no. 4. New York: Carnegie Foundation, 1910.

Freidson, E. *Patients' views of medical practice.* New York: Russell Sage Foundation, 1961.

Freidson, E. *Profession of medicine--A study of the sociology of applied knowledge.* New York: Dodd, Mead, 1970.

Gillette, R.D., Kues, J., Harrigan, J.A., & Franklin, L. Does the family physician's role correspond to the patient's expecations? *Family Medicine*, 1986, 18(2), 68-72.

Glaser, B.G., & Strauss, A.L. *The discovery of grounded theory: Strategies for qualitative research.* Chicago: Aldine, 1967.

Goffman, I. *Asylums: Essays on the social situation of mental patients and other inmates.* Garden City, NY: Anchor Books, 1961. (a)

Goffman, I. *Encounters: Two studies in the sociology of interaction.* New York: Bobbs-Merrill, 1961. (b)

Harvey, L., & Shubat, S. *AMA surveys of physician and public opinion 1986.* Chicago: American Medical Association, 1986.

Haug M., & Lavin, B. *Consumerism in medicine: Challenging physician authority.* Beverly Hills, CA: Sage, 1983.

Haug, M.R., & Sussman, M.B. Professional autonomy and the revolt of the client. *Social Problems*, 1969, 17, 153-161.

Henderson, L.J. The patient and physician as a social system. *New England Journal of Medicine*, 1935, 212, 819-823.

Hyatt, J.D. Perceptions of the family physician by patients and family physicians. *Journal of Family Practice*, 1980, 10, 295-300.

Kelman, H. R. Consumer criteria of health services quality. In E.B. Gallagher (Ed.), *The doctor-patient relationship in the changing health scene.* Department of Health, Education and Welfare, (NIH) Washington, D. C.: U. S. Government Printing Office, 1976.

Parsons, T. *The social system.* Glencoe, Il: Free Press, 1951.

Pratt, L.V. Reshaping the consumer's posture in health care. In E.B. Gallagher (Ed.), *The doctor-patient relationship in the changing health scene.* Department of Health, Education and Welfare, (NIH) Washington, D.C.: U. S. Government Printing Office, 1976.

Reeder, L.G. The patient-client as a consumer: Some observations on the changing professional-client relationship. *Journal of Health and Social Behavior*, 1972, 13(Dec.), 406-412.

Sarason, S.B. *Caring and compassion in clinical practice.* San Francisco: Jossey-Bass. 1985.

Scammon D., & Kennard, L. Improving health care strategy planning through assessment of perceptions of consumers, providers and administrators. *Journal of Health Care Marketing*, 1983, 3(4), 9-17.

Siegel, B.S. *Love, medicine, and miracles.* New York: Harper & Row, 1986.

Starr P. *The social transformation of American medicine.* New York: Basic Books, 1982.

Stephens, G.G. *The intellectual basis of family practice.* Tucson, AZ: Winter Publishing, 1982.

Sussman, M.B., Caplan, E.K., Haug, M.R., & Stern, M.R. *The walking patient: A study in outpatient care.* Cleveland: Western Reserve University Press, 1967.

Szasz, T.S., & Hollender, M.H. A contribution to the philosophy of medicine: The basic models of the doctor-patient relationship. *Archives of Internal Medicine*, 1956, 97, 585-592.

Warner, M. M. The consumer and family physician relationship: Power autonomy, compliance, and negotiation. *Marriage and Family Review*, 1981, 4(1-2), 135-155.

Whyte, W.F. *Learning from the field: A guide from experience.* Beverly Hills, CA: Sage, 1984.

❋ 19 ❋

SOCIAL SUPPORTS AND MORTALITY RATES:
A DISEASE SPECIFIC FORMULATION

Eugene Litwak & Peter Messeri

INTRODUCTION

It is now reasonably well established that social support is an important influence on health. According to recent review articles there is neither a common theoretical definition of social support or a satisfactory explanation of how it operates (Berkman, 1985; House & Kahn, 1985). A major aim of this paper is to fill this theoretical vacuum by use of the task specific theory of group structure (Litwak, 1985) and to present a new classification of causes of death based on this theory. This framework elaborates prior work by designating which causes of death are most likely to be influenced by social supports and which are least likely. It will be demonstrated empirically through an analysis of 1981 mortality data published by the National Center on Health Statistics (1986) as well as a reanalysis of the Alameda data (Berkman & Breslow, 1983).

REVIEW AND REFORMULATION OF LITERATURE ON SOCIAL SUPPORTS AND MORTALITY

Although there is no common theoretical definition of social support, virtually all empirical studies operationally measure social support through primary group ties, that is, family, friends, and neighbors (Berkman, 1985; House & Kahn, 1985). Sometimes investigators also include weak tie groups, such as membership in voluntary associations. On the rare occasions when contacts with staff of formal organizations are included, it is their primary group aspects, such as providing emotional support, and not their official duties,

257

that are measured. Consequently, investigators seeking to explain the workings of social supports are implicitly contrasting them to formal organizations such as doctors and nurses in their professional settings, and not necessarily to the absence of all support.

However, investigators, in their explanations as to how social supports work, do not systematically take into account the role of formal social supports. Therefore, they have not delineated the unique functions of informal social supports. For instance, the literature suggests there are four ways in which social support reduce mortality:

▓ By providing instrumental help,

▓ By providing information,

▓ By providing advice, and

▓ By emotional bonding which buffers stress and/or has direct effects on bodily functions such as blood pressure and the immune system (Berkman, 1985; House & Kahn, 1985).

It is obvious that formal organizations provide the first three services as well. Doctors who give medication provide instrumental help, their diagnosis provides information, and their medical regimen includes advice. As a consequence these three functions do not uniquely differentiate informal from formal supports.

Emotional sustenance would seem one form of support that informal groups are uniquely able to provide. For instance, Berkman (1985) and Cassel (1976) suggest that stress can negatively affect the immune system which leads to illness. Social support, by reducing stress, strengthens the immune system and reduces death. Some argue that expressions of affection, like stroking a pet, can directly lower blood pressure.

Yet, even these explanations do not uniquely indicate when informal group ties are superior to contact with formal organizations. For instance, the adverse health outcomes resulting from the stress of being unemployed can be alleviated by formal organizations which provide jobs. Formal support from psychiatric professionals can reduce stress based on interpersonal conflicts, and physicians can prescribe medication to control blood pressure and boost immune system functioning.

A Task-Specific Theory of Group Structure

Litwak and Figueira (1968) Sussman, (1979) and Litwak (1985) formulated a task specific theory which explicitly sought to differentiate between the types of tasks best managed by either formal organizations or the primary group. Based on an elaboration of organizational contingency theory (Drazin, 1985; Litwak, 1961; Perrow, 1967), the basic premise of this formulation is that tasks can be classified by the same dimensions as groups and that groups can optimally manage those tasks which match it in structure. By looking at the structure of the formal organizations and informal social support groups it is possible to say which types of tasks each can optimally manage.

Weber (1947), one of the early organizational theorists, implicitly assumed that technical knowledge would in most cases enable people to solve problems better than knowledge obtained through everyday socialization. Weber's monocratic bureaucracy was a group structure that was designed to exploit more fully the potential of technical knowledge than the primary group. For instance, members of the monocratic bureaucracy are chosen on the basis of technical training or special job experience, while members of the primary group are chosen because of birth and love, both of which are related to everyday socialization. Members of the monocratic bureaucracy have only a limited commitment based on their technical competence. They can be removed if they become technologically outmoded or do not meet minimum standards of performance. By contrast, the primary group members have long-term, even life-long, commitments. The monocratic organization has a detailed division of labor which permits individuals to train on the job through specialization or alternatively permits task simplification (i.e., taking complicated tasks and breaking them into simpler components so they could be carried out faster and with less errors). By contrast, the primary group is small in size which does not permit a detailed division of labor. Instead, primary group members assume diffuse roles.

To protect technical knowledge from the inroads of favoritism the monocratic bureaucracy maintains an impersonal atmosphere and motivates its members by economic rewards. In contrast, an internal commitment of duty and/or affection and a bartering of services motivate primary group members. These by definition encourage highly personal ties. When Weber, in *The Theory of Social Economic Organizations*, claimed that the monocratic bureaucracy was the most rational or most efficient for the needs of a modern industrial society, he implicitly matched the technical knowledge requirement of tasks with the dimensions of an organizational structure that promoted technical knowledge. Because he assumed that those with technical knowledge would be better able to manage most tasks than those with non-technical knowledge he anticipated that informal social supports would be antiquated in a modern industrial society.

The next generation of organizational theorists demonstrated, contrary to Weber's expectations, that primary-like groups played significant roles in either increasing or decreasing work productivity in complex organizations (Roethlisberger & Dickson, 1939; Shils, 1951; Blau, 1955). However, it was not until the third generation of organizational theorists emerged that more systematic attention was focused on the need to simultaneously classify tasks and organizational structures. The early organizational contingency theorists (Likwak, 1961; Kannon, Coynes, Schaefer, & Lazarus, 1981; Perrow, 1967) recognized that tasks can vary in the amount of technical knowledge and uncertainty. If a task required technical knowledge and was highly predictable then the formal organizations would be optimal. If a modicum of technical knowledge was required and predictability was low then the human relations model would hold. Litwak and Figueira (1968) extended this formulation to its logical conclusion and argued that if little technical knowledge was needed and there was considerable uncertainty then primary groups would be best.

There are, according to Litwak and Figueria (1968), at least two situations where technical knowledge can not be used effectively. First, there are some activities that simply do not require any technical knowledge. For example, a family member with skills gained through everyday socialization is just as well equipped to remove a lit cigarette from the fingers of a sleeping person as a doctor or fireman. A second circumstance where technical knowledge is of little use is where the situation is so unpredictable, or there are so many contingencies, that an expert cannot be brought to bear in time to make a difference, but where a person with everyday knowledge could play some role and is available. For instance, if a tornado destroys a house, starts a raging fire, and leaves one of the inhabitants senseless with a broken leg, a doctor usually cannot be brought to the scene quickly enough to rescue the victim from the flames. A family member at the scene, however, can save the victim's life by pulling him or her to safety. (Of course, had a doctor been available, the likelihood of aggravating the leg injury would have been reduced.) But between the alternatives of a non-available expert and an available non-expert who can provide some help, the latter is more effective. This highlights a very important point which Weber did not fully consider. If a task cannot benefit from technical knowledge, the structure of the primary group may in principle make it faster, less expensive, and motivate people better (Litwak, 1985).

For instance, when technical knowledge is not necessary, the primary group recruitment through everyday socialization is cheaper than the formal organizational procedure because it saves the cost of training experts. In addition, the small size of the primary groups and their lack of a detailed division of labor means they have shorter and therefore faster lines of communication. This is crucial where non-technical services are sought, since large size and a division of labor loses much of its purpose if there is no technical knowledge. Finally, persons motivated by internalized commitments to primary groups rather than the economic incentives of formal organizations are likely to be more conscientious in performing tasks when technical services are not demanded. The chief problem with motivating people by internalized commitments to group members is that it is likely to lead to favoritism.

However, where non-technical tasks are involved there is a minimal risk from use of internalization since, by definition, most people can manage activities requiring only everyday socialization. Moreover, people motivated by impersonal economic incentives require much more supervision than those motivated by internalization. If the work process or outcomes of the former can not be observed, they are likely to do that which is most convenient for them rather than follow the job mandates. Consequently, internalized commitment is better able to insure a motivated individual than impersonal economic incentives when a group has non-technical goals. For these reasons the informal social support group (that is, the primary group) is likely to be less costly, faster, and have a better motivated membership when handling non-technical services.

The Application of Task Specific Theory to Social Supports and Theoretical Convergence

This formulation uniquely differentiates the role of informal social supports from formal organizations in reducing mortality. Thus, informal social supports can provide instrumental aid, information, and advice better than formal organizations for all tasks that do not benefit from technical knowledge and occur in unpredictable situations. By contrast formal organizations are superior for all tasks which benefit from technical knowledge and task simplification.

The theory also predicts the circumstances when emotional support from informal groups is better able than formal organizations to enhance the body's resistance to disease. For example, chronic high blood pressure is a predictable and measurable form of stress on the body that can be successfully controlled through medication prescribed by a physician. On the other hand, daily hassles (Kannon *et al.*, 1981) represent idiosyncratic and unpredictable types of stressors. The primary group members at work and at home are far more likely than a physician or counselor to be available and adept in providing emotional succor to buffer the stress of everyday hassles as they arise. Therefore, the theory predicts that in alleviating the potential adverse physiological impact from the two types of stressors mentioned above, doctors would be optimal for managing chronic high blood pressure, and emotional support from informal groups would be optimal for the hassles of everyday life.

Most recently Umberson (1987) offered a theory of social control to explain the influence of marital and kin ties on health behaviors.

She suggests that the family utilizes social control mechanisms of socialization to internalized health norms or sanctions to maintain them. She, like prior health researchers, does not consider that the formal organizations can also socialize and sanction. To provide the unique social control role of primary groups it would only be necessary to add that they optimally utilize social control in unpredictable situations, where technical knowledge cannot be applied.

A CAUSE OF DEATH CLASSIFICATION BASED ON TASK SPECIFIC THEORY

Application of the task specific theory to mortality studies implies that causes of death can be classified by the extent to which they can be reduced by non-technical or technical services. At one extreme are causes for which prevention of death predominantly depended upon the sorts of nontechnical services best delivered by primary groups. A case in point is when individuals fall asleep at home with a lighted cigarette in their hand. What will greatly enhance prevention of death from this cause is simply the presence in the household of other family members or persons motivated by internalized commitment to chastise the person from smoking in bed or failing that, remove the lit cigarette from the sleeping person. Neither task requires technical training.

In contrast, people who die from Hodgkin's disease illustrate the category of causes which are overwhelmingly dependent on services uniquely provided by formal organizations. This cancer of the lymph glands can only be diagnosed and treated by a doctor.

Prevention of death from a third category of causes requires the services of both primary group and formal organizations. This category is illustrated by a sudden heart attack at home that renders a person unconscious. Surviving such an event might typically involve having a family member call for an ambulance or transport the individual to a nearby hospital for prompt medical attention. If either the primary group member is not present or the formal organization is not close by to provide their respective services the chances of the individual surviving the heart attack would be greatly reduced.

A fourth category of causes of death is that for which group services of any type are of little value. For instance, there is little that others can do at the present time to prevent or treat cancer of the pancreas.

With this classification of deaths in mind, the task specific theory suggests that social support reduces mortality whenever dimensions of the groups match those of the services. For instance, persons high in primary group support but not formal organizational ones will have a reduced risk of dying from causes reduced by everyday socialization (e.g., pulling a cigarette out of the hands of a sleeping person). Those with both formal and informal resources will in addition be less at risk of dying from those causes diminished by both everyday socialization and technical knowledge (e.g., person suffering heart attack at home). Table 1 formally summarizes hypothesized differences in mortality rates derived from the task specific theory. The table corresponds to the four conceptual causes of death categories described above. The rows of the table correspond to population groups that possess different combinations of formal and primary group support or resources. The numbers in the cells represent the hypothesized ordering of mortality rates with 3 equalling highest mortality. To illustrate the tables use, consider rankings in column 1. Prevention of causes of death in this category require services entailing only everyday socialization. Consequently, risk reduction is hypothesized to occur among the persons in the top two rows, who are high on primary group resources (i.e., value of 1). Those in the lower two rows, lacking primary group support, will have high mortality rates (i.e., value of 3).

Table 1.   Hypothesized Rank of Mortality Rates by Formal and Informal Group Resources and Causes of Mortality (Classified by services needed to reduce them)

Group Resources		Causes of Death (classified by technical knowledge and everyday socialization)			
Primary Group	Formal Org.	High Socializ.[a] Low Technic.	High Socializ. High Technic.	Low Socializ. High Technic.	Low Socializ. Low Technic.
		Col. 1	Col.2	Col.3	Col. 4
high	low	1.00[b]	2.00	3.00	3.00
high	high	1.00	1.00	1.00	3.00
low	high	3.00	2.00	1.00	3.00
low	low	3.00	3.00	3.00	3.00

[a]High socializ. means the cause of death can best be reduced by everyday socialization. Low technical means the cause of death can be minimally affected by technical knowledge.

[b]The numbers in the cells indicate the relative height of mortality rates predicted for each cell. A three means the highest mortality rate.

Table 1 can also be used to deduce the relationship between social support and mortality from different "group-service" categories when, as in the case of the data presented below, there are no measures of formal organizational support. This situation is represented in Table 1 by combining rows 1 & 2 and contrasting them with the combination of rows 3 & 4. Table 2 shows the average rankings of these combinations assuming all cells are weighted equally. It suggests a monotonic relationship between causes of death most influenced by everyday socialization and those having primary group resources, even when formal organizational resources are not measured. This logic derives from the fact that rows 2 and 4 in Table 1 are almost constants and therefore, when added to rows 1 and 3, respectively, do not change the direction of the predictions. Since it is often the case that investigators do not have good measures of available organizational resources, this derivation is not trivial.

Table 2.   Hypothesized Rank of Mortality Ratios by Only Informal Group Resources and Causes of Mortality (Classified by type of services necessary to reduce them)

	Causes of Death (classified by technical knowledge and everyday socialization)			
	High Socializ.[a] Low Technic.	High Socializ. High Technic.	Low Socializ. High Technic.	Low Socializ. Low Technic.
Primary Group	Col. 1	Col. 2	Col. 3	Col. 4
Row 1 (1+2)[b] high	1.00	1.50	2.00	3.00
Row 2 (3+4) low	3.00	2.50	2.00	3.00

[a]High socializ. means this cause of death can be reduced through everyday socialization and low technical means that this cause of death can not be reduced through technical knowledge.

[b]The figures in row 1 are an average of the numbers in rows 1 & 2 in table 1 while the figures in row 2 are an average of rows 3 and 4 in table 1.

Tables 1 and 2 also clearly illustrate why previous investigators, who did not distinguish between type of cause or between formal and primary group resources, were still able to show a relationship between informal social supports and overall mortality rates. What they did, in effect, was collapse all the columns in Table 2 into one column. If the columns are combined under the assumption of equal weight, it can be seen that those with high primary group resources will have an average rank of 1.875 as contrasted with the 2.62 for those with low primary group support. This result will logically occur because the high primary group support includes within it those who have both high primary group and high formal organizational resources. People with this combination are most likely to reduce all causes of death. What past investigators did not consider, is the possibility that their findings that informal social supports reduce mortality may have been confounded with the unmeasured effects of formal organizational support.

A six-point scale was developed as an operational version of the above theoretical classification of causes of death. The additional two categories represent intermediate points on an implicit continuum of the level of each type of group service involved in preventing death from a specific cause. In order of the increasing importance of primary group services, the six group-service categories are:

1.   Neither primary group or formal organization services

2.   Overwhelmingly formal

3.   Majority formal/minority primary group

4.   Equally formal and primary group

5.   Majority informal/minority primary group

6.   Overwhelmingly primary group

Two health professionals independently assigned 75 of the largest 3-digit causes of death categories defined in the 9th edition of the International Classification of Diseases (U. S. Department of Health and Human Services, 1980) to the six group-service categories.[1]   The categories had to have at least 1,000 deaths per year and judged to be sufficiently homogenous so that an overall rating could be made. This list constitutes close to 80% of all adult deaths in the U.S. in 1980.   For this study, the group-service categories were ordered to form an interval scale from 1 to 6 as indicated above, showing the increasing importance of informal social supports. When the two raters disagreed on a rating, the cause of death was given the average value of the two assigned categories. This resulted in some of the causes of death in receiving a fractional value (i.e., 1.5, 2.5,. . . 5.5). Ratings for the 75 International Classification of Diseases (ICD) categories are presented in Appendix 1.

We recognize that there are considerable difficulties in attempting to rate causes of death by influence of social factors. For instance, some causes of death have multiple stages with formal and informal groups playing different roles at each stage.   Some primary groups (e.g., the low income ones) may have less knowledge of good health practices than others (Berkman & Breslow, 1983). Some causes of death are influenced by formal organizations while their side effects by primary groups. Before being overwhelmed by these complexities in rating, it should be kept in mind that they may relate to only a few of the causes of death.   It is very important to note that the correlation between the ratings of the two experts was r= .55.   This level of agreement implies that experts can arrive at meaningful ratings, but at the same time, it indicates substantial room for improvement.

THE STRENGTHS AND WEAKNESSES OF NATIONAL MORTALITY STATISTICS

To investigate the validity of both the group-service classification scheme and the predictions of the task specific theory, we have updated and expanded Gove's (1973) analysis of marital status and mortality, using annual mortality data statistics compiled by the National Center for Health Statistics (NCHS, 1986) from death certificates of all people who died in the United States during the year 1981.   Data are available on 8 broad categories of death by marital status, age, gender and race.   For these data we used death rates for seven of these categories: cancer, heart disease, cerebrovascular diseases, chronic obstructive pulmonary diseases, motor vehicle accidents, all other accidents, and suicides.[2]   Homicide was dropped since none of the ICD codes in this category were rated.

Marital status (single/married at time of death) is used as the indicator of informal social support.   Prior work suggests marriage benefits men more than it does women (Berkman & Syme, 1979; Gove, 1973; Helsing, Mysels, & Comstock, 1981; Umberson, 1987; House, Robbins, & Metzner, 1982).   Separate analysis by gender permits some control of this issue.   Other problems with these data are the lack of a measure of health status and limited socioeconomic measures.   Thus there is the possibility that the associations reported below are due to spurious effects of an unmeasured antecedent cause such as poor health (Gove, 1973) rather than the marital bond.   Fortunately, at least four prospective studies have already shown that the effect of social supports on reducing mortality persists even when measures of health and socioeconomic status are introduced (Berkman & Breslow, 1983; House et al., 1982; Blazer, 1982; Schoenbach, Kaplan, Fredman, & Kleinbaum, 1986).

Effects of social supports were estimated from ratios of single to married mortality rates computed separately for each of the seven cause of death categories broken down by age-gender-race groupings of decedents.   Decedents were divided into six age groups: five 10-year groups between 25 and 74 and an open ended category for all deaths of persons 75 and older.   Race was dichotomized into white and Black.   Other racial groups were excluded from this analysis. Denominators for the mortality rates were based on 1980 U.S. census counts for single and married persons in each of the corresponding age by gender by race subgroups.   Mortality ratios greater than 1 indicate that marriage is beneficial, while ratios less than one indicate that the married were more likely to die from a specific cause.

A theoretical score was assigned to each of the seven major

causes of death in averaging the theoretical ratings of the ICD 3-digit categories that define each disease.   The theoretical ratings were taken from the table in appendix 1 and the assignment of ICD categories to each disease category is given in endnote 2.

Marital Status, Age and Race

Table 3 provides the theoretical ratings given to each of the 7 major disease categories and the age standardized mortality ratios for each of the 7 diseases.   The Spearman rank order correlation between them is .93.   This demonstrates that marriage has a strong effect on death rates from those causes that the task specific theory predicts are most amenable to prevention by informal social supports (e.g., suicides and accidents) while matrimony is least beneficial for those causes which are theoretically supposed to be least affected (e.g., cancer and coronary heart disease).

Table 3.   Theoretical Rankings of Seven Selected Causes of Death and Ratios of Single to Married Mortality Ratios - 1981

	Average Theoretical Rankings[a]	Actual Mortality Ratios[b]
Cancer	3.39	1.19
CHD	3.54	1.54
Stroke	3.85	1.63
Pulmonary	4.16	1.96
Auto accidents	4.50	1.85
Other accidents	5.36	2.24
Suicides	5.50	2.17

Spearman rank order correlation .93 (sig. at .01)

aThese rankings are based on clustering the sub set of ICD categories that define each of the seven categories of death and averaging their theoretical scores as given in appendix 1. The ICD categories used to define the seven major causes of death are listed in footnote on p. 12.

bThese are ratios of mortality rates of single persons to married ones. The ratios have been standardized for age, race, and gender.

If the correlations are examined by race and gender some interesting trends emerge.   Categories are similarly ranked across all groups (the Kendall coefficient of concordance, W=.80 p < .01, see Table 4).   However, the rank order correlations are highest for white males (.96), then in order, white females (.89), black females (.54), and least effected are black males (.50).

Table 4. Theoretical Ranks of Seven Major Causes of Death
by Ratio of Single to Married Mortality Rates,
by Gender and Race

Causes of Death[b]	Theoretical Rankings[c]	White Males	White Females	Black Males	Black Females
		Single to Married Mortality Ratios[a]			
Cancer	3.39	1.53	1.36	1.49	1.37
Coronary Heart	3.54	1.96	1.95	1.71	1.68
Stroke	3.85	2.15	1.70	1.76	1.52
Pulmonary disease	4.16	3.05	2.08	2.02	2.16
Auto accidents	4.50	2.63	2.07	1.69	1.44
Other accidents	5.36	3.07	2.72	2.37	2.12
Suicides	5.50	3.22	2.36	1.75	1.93
Spearman R		.96*	.89*	.50	.54

Kendall coefficient of concordance, w= .8000*

[a]This represents mortality rates of single over married.

[b]These are 7 major categories of death designated by the National Center on Health Statistics.

[c]Each major cause consists of a set of ICDs (see footnote on p. 12). The theoretical scores for the general disease categories were computed by averaging the theoretical scores for the ICD codes which defined them (see Appendix 1) assigned to the ICD for each cause.

Another interesting trend in the data has to do with the overall level of the ratios for each gender and race group. As indicated earlier, the larger the ratio the more protective the marriage bond. It is possible for men and women to have the same rank order correlations while the absolute size of the women's ratios can in each category be smaller. This means that both men and women may find that marriage is least likely to protect them against cancer and more likely to protect them against suicide. At the same time it may also be true that married men are more protected from cancer and suicide than married women.

The size of the ratios are compared in Table 5 for each gender and race group as well as for each age and disease category.[3] In 73% of the 42 comparisons, white men were more protected than white women. In 67% of the 39 comparisons, black men were more protected than black women. We also find a race effect. In 93% of the cases white men were more protected than black men, and white women were more protected than black women in 82% of the 39 comparisons. In conclusion, men are more protected by marriage than women, and whites more than blacks.

Table 5.   Mortality Ratios by Group-Service Rating, Gender, Race and Age

Group Service Rating[a]	Age Categories					
	25-34	35-44	45-54	55-64	65-74	75+
White Males						
Cancer	1.42	1.43	1.88	1.45	1.49	1.54
CHD[b]	1.83	1.94	2.32	1.76	1.82	2.08
Stroke	1.56	2.45	2.92	2.02	1.97	1.96
Pulmonary	3.27	3.96	4.15	2.65	2.37	1.89
Auto Acc.	2.37	2.56	3.04	2.61	2.86	2.34
Other Acc.	2.30	2.98	3.85	3.30	3.22	2.80
Suicide	3.21	3.60	4.16	3.02	3.00	2.33
White Females						
Cancer	1.21	1.47	1.42	1.41	1.24	1.43
CHD[b]	2.09	2.19	1.96	1.81	1.51	2.14
Stroke	1.59	1.78	1.61	1.59	1.50	2.10
Pulmonary	2.35	1.80	2.48	2.23	1.83	1.82
Auto Acc.	2.90	2.76	2.28	1.93	1.41	1.14
Other Acc.	2.95	3.95	3.06	2.16	1.76	2.41
Suicide	3.66	3.20	2.29	2.23	1.60	1.17
Black Males						
Cancer	1.14	1.51	1.59	1.56	1.39	1.74
CHD[b]	1.48	1.84	1.64	1.69	1.60	2.01
Stroke	1.55	1.93	1.94	1.63	1.59	1.92
Pulmonary	2.70	1.15	2.73	1.92	1.69	1.93
Auto Acc.	1.38	1.67	1.69	1.63	1.94	1.81
Other Acc.	1.61	2.21	1.42	2.45	2.80	2.71
Suicide	1.76	2.65	1.66	1.42	1.15	1.87
Black Females						
Cancer	1.20	1.47	1.39	1.38	1.36	1.41
CHD[b]	1.56	2.01	1.46	1.57	1.60	1.87
Stroke	0.91	1.63	1.43	1.49	1.67	2.00
Pulmonary	1.79	3.70	2.21	1.82	1.45	1.98
Auto Acc.	1.80	1.88	1.57	1.44	1.02	0.95
Other Acc.	2.10	2.21	2.10	1.84	1.79	2.71
Suicide	1.88	2.88	1.98	1.57	0.47	2.79

The Kendall coefficient of concordance: w=.7974* for white males, .5994* for white females, .4661* for black males, and .5579* for black females.   * sig. .01

[a]These diseases are arranged in the order that they are theoretically rated to be affected by social supports. Cancer is the least affected and suicides is the most.

[b]CHD = Coronary Heart Disease.

DISCUSSION

The rank order correlation between the mortality ratios and the theoretical categories provide confirmation for the task specific theory. The prediction derived from that theory (that causes of death are differentially influenced by social support) calls for an elaboration of House *et al's.* (1982) view that social supports are not disease specific. It may be true that social supports affect all diseases but it is also true that they affect some more than others. In addition, this theoretical formulation offers the investigator an opportunity to explain the differences in mortality that arise because of gender and race rather than treating them as "control" or "background" variables. It directs attention to how gender and race affect the marital units ability to provide non-technical forms of help.

Gender effects. The finding that the marriage provides more protection for males than females is consistent with Gove's (1973) earlier findings using NCHS mortality data from about 1970, Umberson's (1987) work on health behaviors, as well as four prior prospective mortality studies (Berkman & Breslow, 1983; House *et al.*, 1982; Blazer, 1982; Schoenbach, Kaplan, Fredman, & Kleinbaum, 1986). If the task specific theory is correct, the answer to this pattern should lie in the fact that women are better able to manage non-technical aspects of health in a family context. What are the non-technical aspects of health? It means knowledge about early symptoms of illness that leads to early consultations with doctors, it means knowledge of first aid, it means knowledge about preventive behavior such as proper diets, proper weight, need for exercise, etc. It also means that role socialization leads women to be less likely to engage in non-technical aspects of risky behavior such as fast driving, high alcohol consumption, and physical violence.

Once the differences are translated in this way, there are both data and good reason to argue that women are more likely than men to provide these particular types of non-technical health services. For instance, it is plausible that women systematically become more knowledgeable about these non-technical aspects of health behavior because of their role as primary socializers for infants and very young children. These early years require mothers to be in continuous contact with the medical system. Consistent with this hypothesis is Umberson's (1987) data and observation that studies show women are less likely to engage in risk taking behavior (Waldon, 1982), are more responsive to health concerns (Nathanson, 1977) and are more likely to engage in care-giving related to health (Belle, 1982). This means that women are more likely to deliver such health services to their husbands than men are to their wives.

In addition, women tend to be the caretakers of kinship ties, and therefore are more likely to get such non-technical health support from their relatives and friends, than are men (Berkman & Syme, 1979). As a result unmarried women may be less isolated than unmarried men. This would also result in the women's ratio of married to single mortality rate being smaller than men's.

Both of these reasons can account for the fact that marital ties are more beneficial to men than women. Instead of gender being a "background" variable it has been incorporated into the theoretical framework so that the lower ratio of women to men is now viewed as expected.

Race effects. It should be understood that the weaker effects of social supports on black mortality rates was anticipated and as such was more of a confirmation than a rejection of the theory. The raters were confronted with the problem of selecting a reference group to use for their estimates of social support since it was clear that some groups in society had less requisite knowledge of health practices than others and therefore would not use their social support groups as effectively as others. The raters were told to use as their reference group those with at least a high school education. The explicit assumption was made that the ratings would not operate as effectively for poor and less educated groups. Since the NCHS data did not have measures of education or income, the blacks were the only group that could be identified as being poor and less educated.

The question arises as to why socio-economic differences between blacks and whites lead to a lesser protective effect of marriage and how the task specific theory permits the investigation of this problem.[4] To begin with, blacks are likely to receive inferior formal medical services because of their lower socioeconomic standing. The task specific theory suggests that ties to primary groups without access to good formal resources may reduce primary group influence on those causes of death that require services performed by both formal and informal groups. Specifically, this refers to deaths assigned to group-service categories 3.5 to 4.5.[5]

Task specific theory suggests a second possible line of investigation, that is, the impoverished circumstances of many blacks tend to weaken the marital bond even for those who remain married (Liebow, 1967). The weakening of the marital tie is evident in higher divorce and separation rates among the poor than wealthier groups. Even for blacks with viable marital ties, they may benefit less because the poorer and less educated groups in our society are generally less well informed about recent health promotion information (Berkman & Breslow, 1983). A strong marital tie may have less effect on mortality if marriage partners are unaware and therefore discourage healthy behaviors and life styles. For instance, poor families, unaware of the relationship between cholesterol and heart disease may encourage their children to eat eggs and red meat when it is available. The task specific theory, by highlighting the relationship between group structure and non-technical health tasks, illustrates how "race" may be conceived as an indicator for structural variation in informal social supports rather than a mere control variable of little theoretical interest.

Causal direction and longitudinal data. As indicated above, the cross sectional data presented suffer from the inability to determine causal direction which arises, in part, because of the inability to control for health status. This study could only proceed because prior studies have demonstrated that social supports do indeed reduce mortality rates regardless of health status. It is also the case, as indicated above, that the use of marital status to measure social supports is very limited. With that in mind, a reanalysis of the Alameda data, which was both longitudinal and employed a measure of social support that included most primary groups as well as weak tie groups, is important. Berkman and Breslow (1983) seek to show that social supports are not disease specific by demonstrating that there is a social support effect in four major categories of diseases, that is, cancer, ischemic heart, cerebrovascular and other circulatory diseases, and other causes such as suicides, accidents, and pulmonary diseases. However, what they did not comment upon was that social supports had differential impact on these diseases. Cancer was least influenced while that category called "other" (e.g., accidents, suicides, and pulmonary) was most affected. Following the same procedure used with the NCHS data, we classify the ICD categories into these four groupings and assign them their average theoretical ranks as defined in Table 1. We also classified the NCHS data into categories which roughly approximated the four groups designated in the Alameda study. Berkman and Breslow measured the effects of social support by a ratio and compared those with the least social supports to those with the most (See Table 6).

What can be noted in Table 6 is that the rankings of actual mortality rates for the four categories in the Alameda study, the revised groupings in the NCHS study, and the theoretical ratings given to the four major disease categories are the same. This is the case despite the radical differences in measures of social support and the use of longitudinal rather than cross sectional data. In addition, the correlation in the Alameda data is stronger for men than for women, just like the cross sectional NCHS data. For Alameda women, social supports only make a difference between cancer (2.2) and the other 3 categories, and there are virtually no differences between the latter three groups (ischemic heart, 3.2, cerebrovascular 3.2, and other, 3.0). But, for men there is a regular progression of effects of social support as one goes from cancer (1.68), to ischemic heart (2.21), to cerebrovascular (2.25), to the category "other" (3.03).

Table 6. A Comparison Between the Theoretical Rankings of
Mortality Based on NCHS Data and Rankings Based
on the Alameda County Data

Categories of diseases	Theoretical rankings	Alameda County Data	
		Ratio of single mortality rates to married, NCHS for total pop.	Ratio of least social support to most based on index[a] - Alameda
Cancer	3.39 (1)[b]	1.19 (1)[b]	1.80 (1)[b]
Coronary heart disease[c] (Ischemic heart)[d]	3.54 (2)	1.54 (2)	2.44 (2)
Stroke[c] (Cerebrovascular & other circulatory diseases)	3.85 (3)	1.63 (3)	2.52 (3)
Other[d] (Pulmonary[c])	4.16 (4)	1.95 (4)	3.00 (4)
Other[d] (Auto Accidents[c])	4.50 (4)	1.85 (4)	---
Other[d] (Other Accidents[c])	5.36 (4)	2.24 (4)	---
Other[d] (Suicide[c])	5.50 (4)	2.17 (4)	---

[a]This was a complex index consisting of marital status, number of contacts
with relatives and friends, and number of voluntary associations. These data
come from the Alameda county study (Berkman and Breslow 1983). The mortality
ratio from the Alameda study was computed by dividing the mortality rates
for those with the most affiliations into those with the least. By contrast,
the NCHS data divided the mortality rates of married into those of single.

[b]The number in parentheses represents the rank of the ratio with the smaller
number indicating less influence by social supports and the higher number
indicating most influence.

[c]These are the classifications used by NCHS.

[d]These are the classifications used in the Alameda County study and they are
paired with the National Center on Health Statistics classification which
they most closely correspond to. The Alameda study had an open classification which they called "other." The 4 remaining NCHS categories which
were not covered by 3 specific disease categories coded in the Alameda study
were put into this other category but listed separately. In all instances
they exhibited a higher mortality ratio than the 3 specified causes.

Because of the small number of rankings and inability to get
precise matching of disease categories, Table 6 can only be viewed
as suggestive. It does provide some guarded optimism for the
robustness of this theory in dealing with different measures and
longitudinal data sets.

General Conceptual Issues

In addition to indicating which causes of death are most affected by social supports, task specific theory also has implications for more general conceptual problems dealing with social support. One question frequently asked is, "*Do social supports provide a buffering effect or a direct one?*" The buffering hypothesis argues that social supports help individuals cope with stressful situations as they arise. Otherwise, social supports have little effect. The direct effect hypothesis states that social support works during non-crises times as well. Indeed they may prevent the development of crises. The task specific theory states that primary groups can   motivate their members better, be less costly and faster, than formal organizations for managing non-technical tasks.   Insofar as managing non-technical tasks can prevent crises from happening (e.g., providing nutritious meals so people do not get sick, providing emotional support in daily hassles so they do not build up to a crisis situation, etc.), then from a theoretical perspective informal social supports have direct effects.   On the other hand, in crises situations, where unique forms of non-technical help are necessary, the primary group would also be the most efficient group for delivering help so they would have buffering effects as well.   In short, task specific theory provides a unified conceptual framework for Berkman's (1985) important observation that social supports may have both buffering and direct effects.

To make these points clear, consider the stressful situation of a death of a spouse, which also represents a rupture of one of the most important sources of social support for most married individuals (Berkman, 1985; Thoits, 1982).   When this event occurs, other primary group members may act to reduce the resulting stress in one of two ways.   First, they may provide services that were supplied by the departed spouse.   For example, a recent widower may experience stress because he cannot manage the everyday problems of shopping, cooking, laundry, etc., previously managed by his wife.   If his children provide these services, the social support to their widowed father is likely to have a buffering effect.   This is the case because the children's attempts to supply this kind of help while the spouse is alive and well would generally be superfluous or even create friction.   This particular form of the buffering effect occurs because a previously ineffectual primary group (the children) replaces another primary group (marital dyad) that previously provided services.   This buffering effect always assumes a prior direct effect even though it would not show up by the traditional methods used to differentiate the two, because investigators usually only measure the effects of the current support group and not the lost one.   The latter is only viewed as an indicator of a stressful event.

A second kind of buffering effect occurs when the primary group provides a unique form of service which was never supplied by the spouse. For instance, the widower might have an enormous sense of grief for the lost spouse and the children provide him with succor against this grief. The spouse by definition could never have supplied this service. This kind of buffering effect does not require any assumption of a prior direct effect.

What is important to note is that the concepts of buffering and direct effects are low level theoretical observations. They provide little theoretical guidance for stating which groups can substitute for which activities of the shattered primary group or which can provide stress reduction services to an ongoing primary group. Task specific theory does both of these things. Thus, Litwak (1985) points out, by analyzing the structure of a marital unit and the nature of tasks they perform, it is possible to show theoretically and empirically which of the potential support groups (e.g., family, friends, or neighbors) can optimally substitute for the shattered marital dyad as well as which can provide unique services to the intact one.

CONCLUSION

In conclusion, prior writers on social support could not offer a theory of social support which uniquely differentiated social supports from formal organizations. This paper provides a task specific theory which fills this vacuum. When applied to the study of mortality, task specific theory has led to a new classification of causes of death and provided a theoretical rationale for predicting which causes of death will most likely be reduced by informal social supports. This framework and new classification of deaths successfully predicted the rank order in which marital ties reduce 7 major causes of death. Furthermore, this framework provided a theoretical justification for arguing that social supports would provide both buffering and direct effects as well as a new way of looking at these issues. In addition, this formulation enables researchers in health to link their work to non-health related fields that also show the effectiveness of social supports. Finally, this orientation opens up a plethora of exciting and new research paths such as:

* The need to reexamine past longitudinal studies to see if the causal direction of the disease specific role of social supports can be established within a data set which enables one to disentangle the causal effects of health and the ability to look at alternative measures of social support,

* The need to systematically incorporate measures of formal organizations so that the full theory can be tested,

▓ The need to establish a direct relation between the ICD categories and the theoretical ratings rather than working through general categories of diseases so as to see, for instance, if there are some types of cancer that are more subject to the effects of social support than others, and

▓ The need to look at morbidity as well as mortality. ▓

The collaborative efforts of Jack Elinson, Sheila Gorman and Samuel Wolfe in developing the rating schema as well as theoretical framework are gratefully acknowledged. The data for this paper were gathered under funds provided by a National Institute on Aging Grant, #1, R01, AG0 4577-01A1; however, they are not responsible for the analysis or interpretations. We would also like to thank the following for helpful comments on an earlier draft of this paper: Hal Kendig, George C. Meyers, Steven K. Mugford, and John McCallum.

ENDNOTES

1. The two expert raters were Sheila Gorman, a nurse with a Ph.D. in sociomedical sciences, and Samuel Wolfe, a physician with a doctorate of public health.

2. The following are the ICD codes that were used to define each category of death: malignant neoplasm = 140-208; disease of the heart = 390-398, 402, 404-429; cerebrovascular diseases = 430-438; chronic obstructive pulmonary diseases = 490-496; motor vehicle accidents = 800-807, 826-949; suicide = E950-E959.

3. There are some trends which suggest that the correlations between the theoretical rankings and the actual mortality ratios might be weaker for very old women and black males. Since each disease category represents many ICDs it could be that older people may systematically be dying from very different disease categories under the same general rubric. On the other hand, there are some reasons to suspect that social supports may have a "ceiling effect" among the very old which could cause a loss in predictive power. These matters are currently being explored.

4. The lower correlation for blacks may also be caused by the NCHS classifying all people as married even though they are separated. This may be an unusually large group for blacks.

5. A provisional check of the NCHS data tapes where ICD's were classified in terms of the theoretical categories suggests that the biggest differences in mortality ratios between racial groups occur at the primary group extreme (e.g., categories 5 and 5.5) and not in the middle categories (3.5 through 4.5). This suggests that the lack of formal services may not account for the racial differences.

REFERENCES

Blazer, D.G.   Social support and mortality in an elderly community population. *American Journal of Epidemiology*, 1982, 115, 684-694.

Belle, D.   The stress of caring: Women as providers of social support.   In L. Goldberger & S. Breznitz (Eds.), *Handbook of stress: Theoretical and clinical aspects.*   New York: The Free Press, 1982.

Berkman, L.F., & Syme, L.S.   Social networks, host resistance, and mortality: A nine-year follow up study of Alameda County residents. *American Journal of Epidemiology*, 1979, 190(4), 186-204.

Berkman, L.F., & Breslow, L.   *Health and ways of living: The Alameda County studies.* New York: Oxford University Press, 1983.

Berkman, L.F.   The relationship of social networks and social support to   morbidity and mortality.   In S. Cohen & S.L. Syme (Eds.), *Social support and health.*   New York: Academic Press, Inc., 1985.

Blau, P.   *The dynamics of bureaucracy.* Chicago: University of Chicago Press, 1955.

Drazin, R., & Van der Ven, A.H.   Alternative forms of fit in contingency theory. *Administrative Science Quarterly*, 1985, 30(Dec.), 514-539.

Cassel, J.   The contribution of the social environment to host resistance. *American Journal of Epidemiology*, 1976, 104, 107-123.

Gove, W.   Sex, marital status, and mortality.   *American Journal of Sociology*, 1973, 79(1), 45-68.

Gove, W.   Gender differences in mental and physical illness: The effects of fixed roles and nurturant roles.   *Social Science and Medicine*, 1984, 19, 77-91.

Helsing, K.J., Mysels, S.S., & Comstock, G.W.   Factors associated with mortality after widowhood.   *American Journal of Public Health*, 1981, 71(8), 802- 809.

House, J.S., Robbins, C., & Metzner, H.L.   The association of social relationships and activities with mortality: Prospective evidence from the Tecumseh community health study. *American Journal of Epidemiology*, 1982, 116, 123-140.

House, J.S., & Kahn, R.L.   Measures and concepts of social support.   In S. Cohen & S.L. Syme (Eds.), *Social support and health.*   New York: Academic Press, Inc., 1985.

Kannon, A.D., Coyne, J.C., Schaefer, C., & Lazarus, R.S. Comparison of two modes of stress measurement: Daily hassles and uplifts versus major life events. *Journal of Behavioral Medicine,* 1981, 4, 1-39.

Liebow, E. *Tally's corner.* Boston: Little, Brown & Company, 1967.

Litwak, E. Models of bureaucracy which permit conflict. *American Journal of Sociology,* 1961, 67, 177-184.

Litwak, E. *Helping the elderly: Complementary roles of informal networks and formal systems.* New York: Guilford Press, 1985.

Litwak, E., & Figueira, J. Technological innovations and theoretical functions of primary groups and bureaucratic structures. *American Journal of Sociology,* 1968, 73, 468-481.

Nathanson, C.A. Sex roles as variables in preventive health behavior. *Journal of Community Health,* 1977, 3, 142-155.

NCHS. *Vital Statistics of the U.S., 1981, Vol. II, Mortality.* Part A. Washington, DC: Government Printing office, 1986.

Perrow, C. A framework for the comparative analysis of organizations. *American Sociological Review,* 1967, 66, 194-208.

Roethlisberger, F.J., & Dickson, W.J. *Management and the worker.* Cambridge, MA: Harvard University Press, 1939.

Shils, E.A. The study of the primary group. In D. Lerner, & H.D. Lasswell (Eds.), *The sciences: Recent developments in scope and method.* Stanford, CA: Stanford University Press, 1951.

Schoenbach, V.J., Kaplan, B.H., Fredman, L., & Kleinbaum, D.G. Social ties and mortality in Evans County, Georgia. *American Journal of Epidemiology,* 1986, 123, 577-591.

Sussman, M.B. Bureaucracy and the elderly individual: An organizational linkage perspective. In E. Shanas, & M.B. Sussman (Eds.), *Family, bureaucracy, and the elderly.* Durham, N.C.: Duke University Press, 1977.

Thoits, P. Conceptual, methodological, and theoretical problems in studying social support as a buffer against life stress. *Journal of Health and Social Behavior,* 1982, 23, 145-159.

Thompson, J.D. *Organizations in action.* New York: McGraw-Hill, 1967.

Umberson, D. Family status and health behaviors: Social control as a dimension of social integration. *Journal of Health and Social Behavior,* 1987, 28, 306-319.

U. S. Department of Health and Human Services. *International classification of disease* (9th revision) - *Clinical Modifications,* Vol.1 (2nd edition). PHS 80-1260. Washington, D.C. U.S. Government Printing Office, 1980.

Waldon, I. An analysis of causes of sex differences in morbidity and mortality. In W. R. Gove & G.R. Carpenter (Eds.), *The Fundamental connection Between Nature and Nurture.* Lexington, MA: D.C. Heath, 1982.

Weber, M. *The theory of social economic organization.* A.M. Henderson and T. Parsons (Eds. and Trans.). New York: Oxford University Press, 1947.

APPENDIX 1

Group-Service Ratings on 75 of the Largest ICD's

1= Neither formal nor informal, 2= Overwhelmingly formal, 3= Majority formal, minority informal, 4=Equally formal and informal, 5=Majority informal, minority formal, 6= Overwhelmingly informal. Where the raters disagree, an average rank was used, usually giving a midpoint score.

Causes of Death   Ranking

Diseases of heart:

394	Diseases of mitral valve	3.5
396	Diseases of mitral and aortic valves	3.5
402	Hypertensive heart disease	5.0
404	Hypertensive heart and renal disease	5.0
410	Acute myocardial infarction	3.0
411	Other acute and subacute forms of ischemic heart disease	3.0
414	Chronic ischemic heart disease	3.5
415	Acute pulmonary heart disease	3.0
425	Cardiomyopathy	1.5
428	Heart failure	3.0

Malignant neoplasms:

141	Malignant neoplasm of tongue	4.5
146	Malignant neoplasm of oropharynx	4.0
150	Malignant neoplasm of esophagus	4.0
151	Malignant neoplasm of stomach	3.0
153	Malignant neoplasm of colon	3.5
154	Malignant neoplasm of rectum, rectosigmoid junction and anus	3.0
155	Malignant neoplasm of liver and intrahepatic bile ducts	3.0
156	Malignant neoplasm of gallbladder	2.0
157	Malignant neoplasm of pancreas	1.0
161	Malignant neoplasm of larynx	3.0
162	Malignant neoplasm of trachea, bronchus & lung	4.0
170	Malignant neoplasm of bone & articular cartilage	3.0
171	Malignant neoplasm of connective & other soft tissue	3.0
174	Malignant neoplasm of female breast	4.5
175	Malignant neoplasm of male breast	4.0
179	Malignant neoplasm of uterus, part unspecified	3.0
180	Malignant neoplasm of cervis uteri	3.0

Causes of Death	Ranking
183 Malignant neoplasm of ovary & other uterine adnexa	3.0
185 Malignant neoplasm of prostate	3.0
188 Malignant neoplasm of bladder	3.0
189 Malignant neoplasm of kidney & other unspecified urinary organs	3.0
191 Malignant neoplasm of brain	1.0
192 Malignant neoplasm of other & unspecified parts of nervous system	1.0
193 Malignant neoplasm of thyroid gland	3.0
197 Malignant neoplasm of digestive systems	3.0
198 Secondary malignant neoplasm of other specified sites	1.0
200 Lymphosarcoma and reticulosarcoma	3.0
201 Hodgkin's disease	3.0
202 Other malignant neoplasm of lymphoid and histiocytic tissue	3.0
203 Multiple myeloma and immunoproliferative neoplasms	3.0
204 Lymphoid leukemia	3.0
205 Myeloid leukemia	3.0

Cerebrovascular diseases:

430 Subarachnoid hemorrhage	2.0
431 Intracerebral hemorrhage	1.5
433 Occlusion and stenosis of precerebral arteries	4.0
436 Acute but ill-defined cerebrovascular disease	4.0
437 Other and ill-defined cerebrovascular disease	4.0
438 Late effects of cerebrovascular disease	5.0

Accidents and adverse effects:

E812 Other motor vehicle traffic accident involving collision with motor vehicle	4.5
E814 Motor vehicle traffic accident involving collision with pedestrian	4.5
E815 Other motor vehicle traffic accident involving collision on the highway	4.5
E816 Motor vehicle traffic accident due to loss of control, without collision	4.5
E819 Motor vehicle traffic accident of unspecified nature	4.5
E841 Accident to powered aircraft	5.0
E880 Fall on or from stairs or steps	5.5
E884 Fall from one level to another	5.5
E890 Conflagration in private dwelling	5.5
E910 Accidental drowning and submersion	5.5
E911 Inhalation and ingestion of food causing obstruction of respiratory tract or suffocation	5.0

Causes of Death	Ranking
E912 Inhalation and ingestion of other object causing obstruction of respiratory tract or suffocation	5.5
E916 Struck accidentally by falling object	5.5
E922 Accident caused by firearm missile	5.5

Chronic obstructive pulmonary disease:

491 Chronic bronchitis	4.0
492 Emphysema	4.0
493 Asthma	5.0

Pneumonia and influenza:

480 Viral pneumonia	3.5
481 Pneumococcal pneumonia	3.0
482 Other bacterial pneumonia	3.0
485 Bronchopneumonia, organism unspecified	3.5
486 Pneumonia, organism unspecified	3.5
487 Influenza	3.5

Other diseases:

250 Diabetes Mellitus	5.0
571 Chronic liver disease and cirrhosis	5.0
440 Atherosclerosis	5.0
E950- E958 Suicide	5.5

Inter rater reliability R=.55.

CONTRIBUTORS

Margaret Brooks-Terry
Department of Sociology
Baldwin-Wallace College
Berea, OH

Betty E. Cogswell
Department of Family Medicine
University of North Carolina
Chapel Hill, NC

Irwin Deutscher
Department of Sociology
University of Akron
Akron, OH

Elina Haavio-Mannila
Department of Sociology
University of Helsinki
Helsinki, Finland

Marie R. Haug
Center on Aging and Health
Case Western Reserve
Cleveland, OH

Eugene Litwak
School of Public Health &
Department of Sociology
Columbia University
New York, NY

Teresa D. Marciano
Department of Sociology
Fairleigh Dickinson University
Teaneck, NJ

Linda K. Matocha
College of Nursing
University of Delaware
Newark, DE

Peter Messeri
School of Public Health &
Department of Sociology
Columbia University
New York, NY

John Mogey
Department of Sociology
Arizona State University
Tempe, AZ

James W. Ramey
9 Longmeadow Hill Rd.
Brookfield Center, CT

Hyman Rodman
Child Development
and Family Relations
University of North Carolina
Greensboro, NC

Barbara H. Settles
Individual and Family Studies
University of Delaware
Newark, DE

Suzanne K. Steinmetz
Individual and Family Studies
University of Delaware
Newark, DE

Murray A. Straus
Family Research Lab
University of New Hampshire
Durham, NH

Gail G. Whitchurch
Individual and Family Studies
University of Delaware
Newark, DE

Doris Y. Wilkinson
Department of Sociology
University of Kentucky
Lexington, KY

INDEX